A CALL TO ARMS

A CALL TO ARMS

Interlude with the Military

Edmund Ions

We must also subject them to ordeals of toil and
pain and watch for the same qualities there.
And we must observe them when exposed to the
test of yet a third kind of bewitchment. As
people lead colts up to alarming noises to see
whether they are timid, so these young men
must be brought into terrifying situations and
then into scenes of pleasure, which will put
them to severer proof than gold tried in the
furnace.

Plato: *The Republic*, III:413

DAVID & CHARLES : NEWTON ABBOT

The inset photograph which appears on the dust jacket is
reproduced by kind permission of Messrs Marshall of Camberley

Set in ten on thirteen Baskerville
and printed in Great Britain
by W J Holman Limited Dawlish
for David & Charles (Publishers) Limited
South Devon House   Newton Abbot   Devon

Contents

Foreword

In the spring of 1969 I was at a conference of academics at Belfast. The troubles were quiescent, the streets free of skirmish. In any case, the conference was safely housed at the university, well away from the Falls Road. We gathered in the afternoon, pinned small discs to our lapels to show name and provenance, talked, then went back to our rooms and waited for dinner. I looked at the map of Ulster provided by the conference secretary. Some of the names and places were familiar. I had been here before, but that was twenty years ago. Tracing the names along the shores of Belfast Lough and County Down, the notion of a visit took root; a pilgrimage of sorts. The conference diary showed a free afternoon next day. I rang a car hire firm in Belfast.

A day of fitful sunshine, laced by March clouds. I drove east, aiming for the Bangor road, but the new motorway confused me at first, eradicating the landscape I thought I knew, despite the gap of twenty years. Then Palace Barracks came up on the right and memory brought it back. The Twenty-Eighth Training Battalion; harsh crucible; gaunt brick and tarmac. Down there by the Lough, where the single railway line ran along to Bangor, one of the recruits laid his neck across the track on a raw January night. Military police came round company lines next day, saying we were to keep our mouths shut if the press came nosing about.

But there were better memories and I sought them. Driving along the coast I turned off the main road near Helen's Bay. The affluence of the fifties and sixties had sprinkled commuter bungalows along the country lanes, confusing the eye again. Yet there was not much the speculators could do with that promontory above Helen's Bay, and in any case, the War Office has a way of hanging on to its lands. I was certain it would still be there. At Sutra I turned a corner and came upon it: a patch of ground over the sea, with fir trees above brown rocks beyond the clearing. A rusting sign leaned: 'War Department Property'. But there was no gate, and inside, no military presence. Only a crumbling acre of Ireland's 'rank grass and scattered stone'.

The billets had been torn down but the base lines remained among the bramble and gorse. Here and there an end wall half survived, a spot of stubborn plaster clinging. I remembered now that we were the last intake at Helen's Bay camp. I tramped through the weeds, seeking some tangible sign of our presence, expecting none. But suddenly it was there, on a low smooth wall, half hidden among the thorns: a faint patina of ochre, the sand-coloured blanco we scrubbed into our belts and gaiters every evening on this low wall. We had scrubbed hard, fearing demotion to the ranks: officer cadets bound for Eaton Hall or Aldershot or Sandhurst. Not wind nor rain nor ice had quite erased the daily scrubbings where we bulled and blancoed for dog days of drill.

Seeing the faint ochre, some of the voices came back. Genial obscenities in dulcet tones. The coarse participle on Wykehamist lips. Young men with down on their cheeks, fresh from whoring in Belfast, or tumbling the girls in Bangor. Or so they said. Billets were meant for braggadocio: if we were not cruel, we must be licentious. Soldiers could not afford to be less.

I came back to the car and drove from Sutra to Clandeboye; resonant names, drenched in the Irish vessel. We had made camp here among the meadows bordering the woods. We fired blank cartridge and threw thunderflash, hoping to have the smell of battle in our nostrils. The Nissen huts were gone, but among the trees I found a huge tureen, rust-red, holed, mounted on brick. We had 'boiled' our rifle barrels here with scalding water, tugging the wet lint through, four by two, shouting the expletive if the pull-through broke, the sodden lint stuck in the barrel.

I drove to Newtownards. There would just be time, before another slice of history was excavated by the historians at the evening session of the conference. I went south until I came to the wild, tussocked shore at Ballykinler. The firing ranges were still there, facing the sand dunes and the sea, the ranges still in use but silent. The range warden was in his house and said I could go there if I wished, but he gave me a hard look. Dundalk and the Republic were just across the bay.

The butts were just the same. Cold concrete and stiff iron frames. The grey milk of army disinfectant seeping beneath the door of the latrine. Initials scratched with bayonets in the crude wooden benches where we sat, staring up at the targets, waiting for the crack and thump. Frozen days, blowing on our mittens, the sergeant in the middle of the butts, field telephone pressed to his ear, then a stream of expletives along the concrete way as he ordered a shot to be signalled again. When the light faded we hoisted the targets out of the iron sockets and tottered to the storage huts, the huge, top-heavy canvas caught by the sea wind, the corporals snapping at our heels as we stumbled on: a Via Dolorosa through the dunes.

I drove back to Belfast. The conference dispersed next day. Remembrance might have rested there, but curiosity stirred. Returning home I brought a dusty carton from the loft. I untied the rope, noticing the brief code I had scribbled in crayon on top: '5 & 7'. A private joke of sorts. 'Five years with the Colours and seven years with the Reserve.' That was what they told me at the recruiting station when I enlisted in the Black Watch, barely eighteen. My parents were alarmed, but the deed was done. I wished to get away from bookish things for a time. My wishes were respected or indulged.

When I finished the five years, I put away military things, consigned them to this box, putting them out of mind. Now, seeing them again from academic perspectives twenty years later, they seemed distant, detached. Identity discs on a leather thong: name, number and religion for last rites. Army Pay Book, AB64 (Part I). Notebooks from Sandhurst: the Principles of War; Strategy and Tactics. And photographs, some limp, informal; others on stiff cardboard, the trestled tiers rising to the sky as if by divine afflatus. I put them back. Commonplace mementoes, surely; the familiar detritus of anyone who has put on a uniform and slept in a billet. But at the bottom of the carton I found the journal I kept through the five years. It began in a stout black book, ambitiously inscribed 'Persons and Places', and continued through other notebooks, dog-eared from different climates and the vicissitudes of the kit bag.

'Join the army and see the world.' I remembered the poster, and the army had kept its word: the journal confirmed it. Japan, and the mad colonel in the hills, prancing above us at dawn as we crawled on our bellies below, aiming at fitness for battle. Korea, with its silk screen landscape and a dozen summer magpies in a glade: improbable chinoiserie, denying the bric-à-brac of war beneath the trees. Hong Kong, and the gardens at Fan Ling, our mess an elegant villa with lacquer screens: outside, a detonation of flowers beneath the lemon light on terraces and lawns. Singapore, and the night-time cab, cruising between the Mountbatten Club and the Raffles bar; the driver's sudden detour to Lavender Street, anxious to please, the whores climbing in like locusts and we beating them off, fearing an Asian pox. Egypt, and manoeuvres in the desert, the nights awash with stars, the Milky Way a banner above the Negev, and slipping to sleep with a sense of the numinous, dreaming of Zadok the priest and Nathan the prophet. Cyprus, and the walls of Buffavento; and the crumbling monastery at Bellapais, the sightless Gothic eyes staring north over the blue waters of the Levant. Then a slow boat home hugging the shores of Crete up the Adriatic to Trieste, the troopship's tannoy system playing Strauss waltzes, with evening sun on the sugarcube houses of Dalmatia, clinging to the snow-capped peaks of the Dinaric redoubt. Then a train through the black heart of Europe from Trieste to the Hook of Holland: the sad crepuscular of Flanders and Brabant. So back to Aldershot and out.

Certainly I travelled with the army. But the journal evoked the persons as well as the places. A thought occurred. Perhaps a fragment of social history was concealed here; a professional army in the twilight of Empire, its members at odds with the times. The army was already contracting when I joined. Second battalions amalgamated with first battalions in bleak little ceremonies around the shires. Abroad the shadows lengthened over the compounds of the Raj.

Flicking through the journal again, there was much detail. A dinner party at Colombo, for instance. We were flying out to

Korea, a dozen officers, to make up the losses on the Imjin River. We stopped over at an RAF base near Kandy and sipped cocktails on faded rattan chairs, talking with brittle RAF wives in flowered frocks, who fanned themselves and complained of the Labour government. At dusk their husbands scooped them up, vanishing to the bungalows beneath the palms. Ourselves to Colombo to dine in a decaying hotel at a round table, the napery limp, the damp, coffee-coloured fans stropping ceiling shadows. We talked of loose women and a major sent the curry back. Not strong enough, he said: he'd been at Rawalpindi before the war. He belched, his fist descending on a pyre of popadoms. We drank cold lager from green pearled bottles, ate our way through the molten curry, flung notes on the brass tray, and tottered through a fronded hall. We boarded the lorry and sang lewd songs along the road to Kandy, the scent of jacaranda on the night air. The lights in peasant hovels winked on, then off. But time was on their side.

That could be one perspective on the military, I noted, but through the journal another emerged. The stereotoype was much too easy. Serving soldiers are taciturn men, certainly not given to public utterance. When they do speak out, it usually lands them in trouble. Servants to politicians, they pick up the pieces when others have made a mess. During the fifties they did not have it so good. Abroad, they accepted conditions in dust-dried barracks that would have brought shop stewards to apoplexy. During the sixties they became the butt of satirists and cinéastes: rich young men who have done well out of the peace. I think they deserved better than that.

There were further reasons for a memoir, therefore. If I was not exactly of the army during the years I spent with these men, nevertheless I was with them and saw their lives. If they were short on original ideas and lacked imagination, they had other qualities easy to overlook in the age of affluence. Unselfishness, for instance; and a personal code of honour; a stubborn integrity; a rooted tendency to keep their word. Unfashionable qualities, now in short supply, not over-abundant among smooth men in the media.

Many of the soldiers and men I knew were compulsorily retired, caught in the web of political change as Britain receded from a world role. None of them would retire rich: the army made sure of that. Others are dead. Some died young. If the prime reason for this memoir is to sketch a fragment of social history, therefore, a secondary one is to qualify the crude stereotype of those not disposed to defend themselves with the pen, if only because they are too busy defending those who can and do.

Two problems presented themselves for a faithful account. One concerns utterance. It is no secret that among the military the four-letter word is a permanent obbligato, salting every phrase. Current literary fashion favours its inclusion, arguing, it seems, that the four-letter word is honest Anglo-Saxon, familiar to all of us. That may be so, but among the military the word connotes rapine, not love; woman a mere orifice for the male member: a quick, off-duty poke, then back to the billets to boast. Again, a word over-used is a word without meaning. I have therefore excluded four-letter words and their variants, except for those occasions where the particular context requires its special savour. But let the avant garde be assured that as a soldier I used the word as often as they could wish. Or oftener.

The second problem relates to persons. A journal is a private thing, and young men sow wild oats. There is a vogue for exposure these days, confusing it with candour. Yet a memoir is not necessarily the confessional, and privacy is what civilisation is mostly about. I have therefore excised the details of unmilitary friendships which were private compacts and must remain so. Except one, and that by permission, because of its special intensity, its contrapuntal quality, and because it kept me going when the tedium of military life was severe. Elsewhere I have omitted names or screened identities wherever I felt this might remotely be wished.

I must record a debt to Colonel G. A. Shepperd, librarian of the Royal Military Academy, and to John Keegan, an Oxford contemporary, now a senior lecturer at Sandhurst. Both gave generously of their time when I made a visit to Sandhurst to discover memory of things past.

CHAPTER 1

Enlisting

The recruiting station occupied one half of a long building, converted from a garage soon after the war. I passed it often enough on the bus when I went into Newcastle to study at the reference library. The other half of the building was an ambulance station. The irony of the juxtaposition did not occur to my schoolboy's mind. Up to now I had looked down at both from the upper deck of the bus. The little offices were mere partitions, separated by hardboard. It was late autumn of 1947 and the age of austerity showed its marks everywhere. Drab clothes, a greyness to life, untenanted shops, and here and there forgotten sandbags revetting hidden windows.

I went in where a small notice said 'New enlistments'. A narrow passage showed various painted signs outside the office, sticking from the walls like tiny jetties from a straight shore. I did not understand the initials on the signs. Six months later I could translate 'DAQMG', 'RQMS', and other devices by which military men abbreviate their rank or station—but at this moment 'CSM' was the limit of my knowledge. Fortunately, a man in uniform appeared from one of the offices. He raised his eyebrows, and said 'Lost?' in a friendly tone which had none of the ferocity my schoolboy fears led me to expect. He was an enormous man, dressed in blue patrols, and I suddenly remembered that I had seen him now and then striding across Barrass Bridge in the city centre. He was a sergeant in the Guards. His uniform seemed to be sewn on to his limbs.

'I've come to join up,' I said, wondering if the idiom was correct. 'Come to enlist? Over there. Second door on the left.' He smiled again, but seemed surprised and gave me a close look as he passed. I knocked at the door marked 'Inquiries'. Nothing happened for a minute or so, though I could hear stirrings inside. Then someone opened the door—a youth of about my age.

'We're waiting. He's not back yet.'

I went in and sat down next to the youth. Two others were alongside him. In the minutes that passed, each of us took stock of the others. I seemed to be odd man out. I was wearing a grey suit, a collar and tie. They wore open-neck shirts, their hair was lank, fairly long; their double-breasted jackets and old flannels, I was embarrassed to notice, seemed to place them from parts of the city where grammar school neatness would necessarily be rare. They seemed to be my age, but their shoulders and their hands showed that manual work, not book learning, was their common experience. No one spoke. I took up one of the tattered tabloids on the spare folding chair and pretended to be absorbed. In reality I was wondering whether my schoolmasters and my parents were not right after all: that I was making a disastrous mistake.

I was just eighteen. My decision to spend five years in the army had crystallised slowly, and in a complicated fashion. Looking back on it with the ease of hindsight, I see some aspects that stand out clearly enough. I had worked intensely hard at school. During my final six terms I had taken it as a sign of personal failure if I was not top of the 'A' form in both the term's marks and end of term examinations. The distinctions and credits in Matriculation examinations were chiefly in the humanities. The History, English and French results prompted some of the masters to urge me to try for an Oxford scholarship. But possibly I had worked too hard. Certainly I felt stale, intellectually constipated. Another three years of book learning, of writing more essays on Metternich or Bismarck or the Eastern Question seemed an unpleasant prospect. I declined to try, though to my history master I said 'Perhaps later.'

My parents had remonstrated. Surely National Service would give me a long break from studies. I would be conscripted any day now. Indeed, I had already had my medical tests. Why enlist for five years when conscription would claim me for two years in any case? My ignorance of military life was total; so much I was

forced to admit. There was no OTC at school and I should have
been the last to join if there had been. I had never joined any
form of army cadet force, and was repelled by the few glimpses I
had caught of them: muscular youths in badly fitting uniforms,
on the march behind bands, with forage caps perched on greasy
spiv's hair. This was schoolboy prejudice, of course. I saw what
I wanted to see. But I was not a joiner of anything: much more
of a solitary. I had never joined the Boy Scouts. The Boys' Brigade, with their raucous bugles, bell-boy hats and black uniforms
had always conjured unpleasant associations.

My reasons for volunteering for the army were complex, therefore. From the skein of tangled motives I can recall some. At
school my French master was a scholarly man, more perceptive
than the other masters, and he saw the danger signs as term marks
and examinations continuously placed me first in the lists. A
touch of arrogance had come. He took me aside quietly and
warned me. Hubris. Insolent pride. They might herald a fall. I
was wounded, but knew he was right. It had become too easy,
even though I always worked hard. Perhaps I should test myself
—a thorough test. Briefly, romantically, I thought of the French
Foreign Legion. I looked up some books and memoirs and
dropped the idea. The Beau Geste myth evaporated among the
thugs, the swindlers and murderers who seemed to make up the
column.

For some time I had also been reading a spate of biographies,
devouring two or three a week. I knew that Gibbon had been a
Hampshire Grenadier, and none the worse for that. Tolstoy,
Schiller, Coleridge, Churchill, Thucydides: disparate men, yet
each had experienced, or endured, a military education of sorts,
and they seemed the better for it or, if not, it had not stultified
them. Again, Plato had argued the virtues of military asceticism,
crucible for tempering youthful alloys into stronger stuff. If Plato
said so it must be true. I was seventeen and impressed by big
names.

I consulted my brother who had been called up near the end of
the war and who had fought in Normandy and Malaya with the
Cameron Highlanders. We walked and talked as my decision crys-

tallised. He suggested the Black Watch: the Royal Highland Regiment. Their tartan was a subdued combination of dark green and black. The regiment's motto was 'Nemo Me Impune Lacessit.' He added an earthy translation to the correct one. I decided on the Black Watch.

My parents' alarm at a five-year enlistment became quiescent, now that my brother's assurances were added. It would be good for me. I would travel. Millions of men and women had done the same during the war, and were all the better for it. There was no war now, but the travel and the experience were still there for the asking. My mind was on the travel and the experience: I overlooked the tedium of dusty billets and fly-blown camps, oven-hot under the sun.

But as I sat in the recruiting station apprehension overlaid all else. I could hardly back out now, and once my signature was on the forms I had filled out at home (but not yet signed, as though hoping for some sudden veto from above), it would be too late; I would be a signed-up soldier, subject to military law.

The door opened and a sergeant came in. Sergeants had three stripes, that much I knew. He scanned us rapidly as he went to his desk. 'New enlistments?' he asked. We nodded. 'And all had your medicals?' We nodded again. 'Good. It won't take long.' It sounded like the reassurance of a dentist, or a surgeon about to remove a limb. He took in the forms we handed over, then checked the particulars with each of us in turn. Date and place of birth; father's occupation; schooling; any jobs held since school; results of medical tests.

My turn came last. The others had worked at a variety of jobs —chiefly casual labour on building sites. I had to admit that my schooldays were barely over. When the details were checked we each signed the form. Then the sergeant took up a Bible and asked us to stand up. Each of us placed his hand on the Bible and repeated the oath of allegiance to King and Country. The sergeant put down the Bible and said, 'That's it. You're in the army now.'

My brother had told me I would receive the King's Shilling after signing on, but this did not materialise. Perhaps the tradition had been cancelled during the post-war austerity, or the sergeant had overlooked our bounty. I thought it best not to ask. The three others filed towards the door and I was about to follow, but the sergeant asked me to stay behind. When the door was shut he took up the form I had signed.

'Are you quite sure you want to join the army?'

'Yes, sir.'

'It's "Yes, *sergeant.*" You call officers "Sir". '

'Yes, sergeant.'

'You've got all these school certificate results: matriculation certificates. Haven't you thought about doing something else; going to university?' I said I had thought about it, but didn't want to just yet. I wanted a change.

'Wait here,' he said, and went out, taking my signed form with him. I wondered if the army was about to refuse me as unsuitable material. Presumably they were entitled to do that. I was still surprised that there had been no interview before the signing on ceremony.

The sergeant returned with an older man whom I took to be an officer, though I was uncertain about his badges of rank. He wore a tunic and a shiny leather belt, and had some insignia on his epaulettes; and he spoke with a different accent. The officer took the sergeant's seat at the desk and the sergeant stood to one side, his hands clasped behind his back.

'You've chosen a good regiment,' the officer said. 'The Black Watch have a great record. Are any of your family in the army?'

I said my brother had served in the Cameron Highlanders during the war, but no one else in the family had ever served with the military.

'No family tradition, then,' the officer said, which seemed unfair. There was a strong family strain of individualism, of Northumbrian independence, of self-reliance. Another silence followed. Then the officer said, 'We've decided to send you for basic training to the Twenty-Eighth Training Battalion.' He paused. 'Have you heard of that?' 'No, sir.'

'It's for potential leaders. Officers and warrant officers. Both. The training is much harder. It's bloody hard, in fact. Do you

B

think you could stand it?' I gave the expected reply. 'Good,' he
went on. 'You're A1 for medicals, I see, so you ought to manage
it. You can report in January. That's the next intake. It's in
Ulster. Near Belfast. Ever been to Ireland?'

I shook my head. 'No, sir,' I said.

'Good. Well then, that's settled. Good luck!' He held out his
hand and I shook it. The sergeant held his out too. 'Good luck,'
he said.

Outside in the sharp air of an October afternoon I wondered
what I had done. But there was also the rather heady feeling of
having taken a difficult decision long delayed.

CHAPTER 2

Palace Barracks

My posting instructions arrived a few days later, including my army number and a rail warrant for a one way journey to Belfast by train and ferry. The next intake at the Training Battalion would begin on 12 January of the New Year. If I wished to arrive earlier, I was to let the adjutant know at Palace Barracks, Holywood, near Belfast. I thought it might create a favourable impression on the military if I showed that I was willing to enter the army a few days before my training began, so I wrote to say that I would arrive on the first Monday of the New Year, 5 January. My offer turned out to be misguided, but on the Sunday evening when I set out from home with the minimum of luggage —as the posting order crisply required—I could hardly know this.

To arrive at Belfast fairly early on Monday I had to travel overnight. The Stranraer train was waiting when we arrived at Newcastle station and I shook hands with my father on the platform just before midnight. He was still slightly puzzled, but all discussion about my decision had ceased weeks earlier. The independence I had claimed was, I now knew, accepted and respected. There was certainly no going back.

The train was crowded with servicemen—a mixture of army, navy and air force. We drew out into the night and rolled along the Tyne valley where Hadrian had flung his wall. At Carlisle there was a break of half an hour whilst the boat train awaited a connection coming north from Crewe. The bar and buffet on the platform were like part of a wartime film, with soldiers in forage caps clustered about a tea urn among a litter of cracked cups and stale buns.

The next train took us away to the west in the night, through the fields and valleys of Kirkcudbright, under chill January

moonlight. At Stranraer we stumbled from the brown capsule of
the train and showed our travel warrants at the ship's side under
bare electric lights. It was bitterly cold. The ferry slipped away
around four in the morning and we headed for Ulster on a
choppy sea. I fell asleep on a bench in the warm stench of the
main lounge.

At Belfast I had breakfast in a café by the quayside then found
the bus station. Shortly after nine in the morning I was put down,
alone, at the gates of Palace Barracks at Holywood, a few miles
along the shores of Belfast Lough. A high spiked iron fence surrounded gaunt brick buildings. I thought of prison. Perhaps
'Palace' was some sadistic military joke. Later I discovered it referred to the Bishop's Palace, once on this same spot. The Protes-
tant Bishops of Downe, Connor and Dromore.

A tarmac path led up a slight incline to a sentry box, its roof
pointed and crenellated like a child's toy. A soldier was standing
there with his rifle butt resting on the ground, his feet apart. He
seemed to be ignoring his surroundings. I wondered whether you
were allowed to ask for directions from a sentry on duty. I walked
up and said that I had come to report. His eyes swivelled and he
cocked his head. 'Guardroom over there,' he said curtly. The
brick building was a few yards ahead. I thanked the sentry and
noticed that the creases in his uniform were knife-edged, his toe
caps like bright enamel, his hair very short behind his beret.

I knocked at the door marked Guardroom. A stamping noise,
and then the door was flung open violently. A man with two
stripes on his arm and a white belt said, 'What d'you want?' I
explained my business. 'New intake,' he said. 'That's not till next
week. Baker Company lines you want.' He drew me to a map on
the wall. 'You're here,' he said, and tapped the map with a tea-
spoon from the trestle table, where five or six pint-sized mugs
were gathered in a cluster. 'There's Baker Company lines, up
there. Beyond the square. Past the clock tower. Got it?' He prodded the map once more.

I found the clock tower by the side of the square. I felt alien
among various groups marching stiffly, like toy soldiers, in different directions. My plain clothes marked me out. Baker Company

lines lay beyond the broad asphalt square, which seemed about an acre in extent, so I set out across it on a diagonal route. This was a mistake. I had gone about fifteen paces when an extraordinary noise came from somewhere near a line of billets at the other side of the square. I stood still for a moment, and the noise reached me much more distinctly. '*YOU* there! Get OFF my bloody square. Get OFF!' The source of the noise was clearly a bulky uniformed figure in a peaked cap with a stick beneath his armpit. I sensed that he had a good deal of authority, especially on or about the barrack square, so I stopped. 'Come HERE,' he shouted. The command was very distinct. I walked quickly towards him. 'DOUBLE,' he shouted now, and instinct made me run, my small suitcase waggling at my side. I put my heels together as I stood before him.

'What do you think you're ON?' he said. Curiously, he spoke quietly now, very distinctly and in measured tones. I learned later that this was merely the prelude before the storm, the adagio before the fugue, *con brio.* I said I had been directed to Baker Company lines, beyond the square. I injected several 'sirs' into my explanation, as I had already noticed a large motif sewn on his sleeves at the biceps, suggesting rampant lions, with possibly a royal motto. His cap had a bright, sharp, shiny peak which almost hid his eyes, so that he had to tilt his head backwards slightly to see me, despite his height. 'So you're new intake are you?' he said, taking the stick from beneath his armpit and swinging it a little before him, the bright brass ferrules coming close to my knees. 'Well then, your first lesson in the British army is that you never, NEVER walk across this square. Because this square is SACRED GROUND, UNDERSTAND?' He barked very loudly now.

'Yes sir,' I said, quietly, meekly. I was still uncertain whether he had not some statutory right to hit me with the pointed stick.

'So just you get off my square right NOW, SHARP, and don't ever set foot on it again, 'less it's squad drill. UNDERSTAND?' I said 'Yes sir' again and went off quickly. He was not finished, however. I learned later that they rarely are. 'And get your HAIR cut, TODAY,' he bellowed. I shouted 'Sir!' and got away from

the RSM's domain as quickly as I could. It was upsetting to be
told to have yet another haircut. I had asked for a specially close
cut at home only two days earlier. But now I noticed that the
young men marching past in squads all showed whiteness of the
scalp behind the ears and beneath the beret. Once more a vision
of prison life came to mind.

I found Baker Company office by a circuitous route. Inside, a
soldier with one stripe on his sleeve was typing at a decrepit
machine. I stated my business, then learned that my early arrival
was more likely to create problems than to be welcomed.

'You'd better see the company sarn major,' the lance-corporal
said, and vanished through a connecting door beyond the round
iron stove. A concealed hatch slid open near my ear and a face
appeared. 'Come through,' it said.

The sergeant major was in fact a friendly man. There was no
sign of the red face, moustache and puttees which the cartoons
had led me to expect. Instead, a man in his forties with careworn
face listened to my story as the lance-corporal stood by. The sergeant major remembered a brief memo from the adjutant's office,
and he took it out of his in-tray to confirm the details.

'That's right, you're arriving today,' he said redundantly, but
emphatically. 'Well, nothing will happen until the intake arrives
next Monday, so you'll find yourself on fatigues most of the time.'
I supposed this was meant to be a form of punishment for having
arrived early.

'You'd better draw a bed and some denims and get out of those
things,' he went on. 'We'll get you kitted out tomorrow. Better
go to the barber this morning. That's the first thing.'

Suddenly I found a little courage. 'I had it cut on Saturday, sir.
I asked them to cut it short.'

'Turn around.' I turned and stood still.

'It's not a Palace Barracks haircut,' the sergeant major said.
'Best get it done. The RSM likes short hair. Corporal, take Private Ions and draw one bed, bedding, one pair denims. Take him
to number three billet, when you've got that.'

In the company store a sergeant with a crown on his sleeve
appeared from behind a mound of blankets. He looked older

than the sergeant major and shifty. His eyes were never still.

'CQ, this man wants bed, bedding, denims. New intake.'

The CQMS indicated a stack of folding iron bedsteads against the wall, and with the corporal's help I disengaged one. It was heavy, buckled, and it clanked. Meanwhile the CQ placed on his short counter—a tiny chasm between mounds of bedding, cans, tripods, and ammunition boxes—one set of denims, the mole grey suiting I had seen on the recruits about the camp. He counted out some bone buttons and an equal number of rings. Three blankets were flung down from a tall pile beyond the counter. A small dust cloud rose.

'Youse'll be needing the biscuits too,' the CQ said, in an Irish brogue. I supposed he was about to provide my first meal in the army, but instead three flat cushions making up a lumpy mattress descended from another pile.

'That's what you sleep on,' the lance-corporal said helpfully, patting the biscuits and raising more dust. Then a pint-sized mug, a segmented aluminium tray, a knife, fork and spoon were pushed towards me.

'Right,' said the CQ, producing a limp book discoloured by the rings of tea mugs. He scribbled in the entries, swivelled the book about expertly. I took the pencil stub and signed.

The corporal showed me to number three billet in the 'spider' of six wooden billets connecting to a central ablutions. We carried the bed like a stretcher, with my newly drawn kit spread upon it. The corporal tramped up two steps to a battered door, flung it open, and said, 'In here.' The billet was completely bare from end to end. A startling sight. I suddenly realised how much one takes furniture and carpets for granted. There were no curtains.

'Where you like,' the corporal said. 'But take my tip. Far wall. Keep well away from the door here. You'll get your kit nicked. And the drunks. Sick on top of you. Nearest to the door always gets it worst. Keep well away.'

He helped me to the end of the billet and I set my bed against the end wall. The corporal pulled it back a yard. 'Leave a space. Inspecting officer has to get round your bed, see your kit. Regulation distance three feet. Line up on that line.' He indicated a white line painted the length of the billet and again tugged the feet of the bed so that they rested square in the middle of the line.

'Know where the NAAFI is?'

'No, corporal.'

'Down there, turn left, past the cook-house, left again. Can't miss it.' I thanked him and he prepared to go. 'Best make up your bed. Get into denims. We'll kit you out tomorrow. The barber's next to the clock tower.' He left.

I sat on the bed and turned the denims over. Putting on the bone buttons presented no problems, though some of the rings were stiff and unyielding. I wondered whether I was allowed to wear my own shirt—my own anything—underneath the denims. A rehearsal, wearing nothing beneath, persuaded me to take a risk, whatever the rules required. I buttoned the denim tunic close up to the neck, but the brass hooks were slack, and the neck itself much too large. In fact the whole outfit seemed baggy and unkempt. My white shirt protruded well above the neckline. Perhaps on my first day ignorance of regulations could be pleaded in mitigation. I folded away my suit and my tie in the suitcase and sat on the bed.

It was not at all clear what was now expected of me. I was alone in an empty billet surrounded by space. It was cold. The billet was unheated. The company lines were deserted. The best thing seemed to be to have my hair cut. I discovered a route which took me well away from the barrack square. I heard later that the RSM spent most of his day hovering in the shadows of a huge chestnut tree at the edge of the square, guarding his asphalt like some mythic monster standing guard over a holy grail. One story had it that the RSM did not need sleep, and that his wraith-like figure could be seen in the shadows like Banquo's ghost at all hours of the night.

The barber's shop was strictly functional. The assistants wore khaki overalls and worked swiftly. A sprinkling of recruits sprawled on folding chairs. My turn came soon enough. I did not expect any polite inquiry on how I would like it cut, and none came. The electric clippers travelled straight up, from the nape of the neck to the crown, in a series of vertical furrows. My scalp felt very cold at the back. A few snips with what seemed to be tailor's shears, then the grey cloth was whipped from my neck. I caught a quick glimpse of myself in the mirror, hid my sense of alarm, tried to look matter of fact as I left. There was no charge, no tipping; a three-minute encounter.

It was now mid-morning. Back in the billet I was still unclear what was expected of me. The corporal had mentioned the NAAFI, but whether that was an all-day restaurant, an off-duty bar, or a shop, or all three, I was not sure. I realised again that my ignorance of military life was almost total. To go there might be some privilege reserved for recruits in training. Moreover, an army barracks was clearly no place for an idle stroll. I preferred to avoid another clash with authority, so I decided to lie low in a literal sense. I took out the book I had brought and stretched out gingerly on the bed. The springs were spongy and set up wave-like effects.

I must have had about half an hour's undisturbed reading, punctuated by distant noises of drill on the main square, when the billet door was flung open. I was coming to realise that in the army all persons of authority tended to fling open any door har-bouring those of inferior rank. Possibly it was meant as an un-spoken warning signal. A shot across the bows. A corporal stood at the open door. He wore a red sash across his chest and he carried a small, knout-like stick.

'What the hell are YOU on?' he said very loudly. Instinct prompted me to get to my feet quickly. I began my explanations.

'Stand to attention,' he said, approaching. I put my feet to-gether, though once more it seemed a ridiculous stance from which to start a conversation, or any verbal exchange.

The corporal may have been put out slightly by the novelty of the situation. I was obviously in the army, but in some sense not yet of the army. He circled my bed silently, and prodded the blankets, heaped to form a pillow.

'I'm orderly corporal,' he announced, and touched the red sash across his chest. 'One thing you'd better learn. Don't let anybody catch you lying on your bed during parade hours. Against Standing Orders. Got it?' I said I got it.

'Right. First thing you do is tidy up your bed. Box your blankets. Then I'll give you a job in the area. Know how to box your blankets?'

The term was foreign to me. I said, 'No, corporal, I'm sorry.'

'Well, come here.' He looked more contented suddenly. No abstractions. A practical problem to be solved. He laid down the short stick he was carrying and with relish flung the blankets across the two biscuits near the foot of the bed. 'Hold that end,' he commanded. 'Now fold. I'm helping you now, remember, but tomorrow you do it yourself.'

We folded two of the blankets into four. The corporal nipped the edges and patted them about to form two rectangles. He placed one on top of the other, and pointed to the third.

'Now wrap it round the other two and tuck in. *You* do it.' He put his hands on his hips and watched my efforts for a minute or more.

'Come here,' he said, and swept me aside. With a few deft movements he wrapped the blanket about the other two, tucked in the overlap, nipped the corners of the envelope into a box, lifted the whole contraption, turned it over so that no tucks or overlap showed, and placed it carefully at the head of the bed. He centred the construction, nipped and tucked a little more, and stood back reverently.

'That's boxing your blankets,' he said. 'In the Twenty-Eighth Training Battalion you box your blankets. At reveille,' he added. 'Now come outside.'

He walked me along the outside of the billets gesturing here and there.

'See those bits of paper? I want this area cleaned. Every little bit. Waste bins are back of company office. Over there.' He pointed with his stick once more. 'Now move,' he said.

I advanced on a toffee paper. The company lines bordered a soccer pitch, and a certain amount of spectators' litter had blown across the iron hard ground. I gathered an assortment of chocolate wrappings, sweet papers, and scraps of newspaper. (Did the army employ people especially to shred newspapers for the induction of new recruits?) I thought a good deal about my experiences during the day. My present job was not so much ignominious; more an inefficient use of manpower. I wondered why the army did not have pointed sticks, like the park attendants in *Punch* cartoons. Was this what was meant by 'fatigues'? I was not sure what standard was expected. Was every tiniest speck of foreign matter to be removed? I made some limited discoveries about the grubby treasure hidden about the common plot. Buttons, bones, nails, small pieces of webbing, rusty boot polish tins, torch batteries, used condoms.

After about an hour of this, a bugle sounded from the direction of the barrack square. The lance-corporal from the company office came up and said, 'That's cook-house. Best get there quickly before the queues. Take your eating irons, your mug and tray.'

I disposed of the final two handfuls of rubbish, rubbed my hands under icy water in the ablutions, and went off to the cook-house. Half way there I heard the crisp noise of boots overtaking me. The sound incorporated authority once more. I walked quickly, holding my tray under one arm and swinging the other rather manically, my finger crooked in the handle of the pint mug. I was overtaken and passed, but as he passed, a tall sergeant with a red plume on his beret said sharply, 'Button up!' I thought on the instant that he had X-ray eyes, and my hand searched quickly at my fly. He turned his head and barked, 'Rear trouser buttons!' Neither of my rear pockets was buttoned. 'Sir!' I called out, and put the matter right. I was already struck by the economy of language in the army. It had some virtue.

The cook-house and dining hall were again strictly functional. Bare trestle tables, no curtains, bare floors. I joined a queue and moved forward, listening to the exchanges between recruits already in training. They looked remarkably healthy and in fact very cheerful. I heard accents which lacked the various regional inflexions I had met during the morning. Instead these voices were distinctly upper-class. Looking now at some of the recruits I saw faces which, if the hair had not been cut short and if the denims were wished away, belonged to fox-hunting and country house environments. It was all the odder, therefore, to hear the repetition of the crude expletive—as it then seemed to me: the four-letter word which I had heard several times that morning, chiefly from the orderly corporal as he initiated me into the boxing of blankets. This same expletive—used adjectivally most of the time, and sprinkled haphazardly about their sentences— sounded bizarre from these fresh-faced young men in the cookhouse queue. They used it unthinkingly, as though it had lost any meaning or force. Eventually I came to see that this was the case. It did not seem to me an attractive word, whatever the context, but one borrows language to assist easy communication, and within a couple of weeks I began to adopt the common language of the group.

Lunch was served by a line of ill-looking young soldiers, greyfaced, as though they rarely saw daylight. They filled the various segments of the aluminium trays from big ladles. The meal was ample, but extremely plain. Not much delicacy accompanied the ladling process. Custard mingled freely with gravy in adjacent segments of one's tray. At the end of the hotplate a sullen looking cook dipped a pint mug into a vast urn of steaming tea and emptied it into the outstretched mugs of the recruits filing past. The tea looked grey and weak; the serving mug was cracked and stained.

I sat at an empty table, anxious to avoid another round of explanations about my arrival that day. A small group occupied one end of the bare board as I sat. They talked mostly about NAAFI girls and ignored me as they ate their meal feverishly, waving their knives to stress a point. I noticed an officer strolling slowly from table to table, with the orderly corporal in close attendance. The officer bent down here and there to have a word with the recruits. Now and then laughter erupted and the officer smiled and gripped his stick with both hands behind his back as he cruised on.

I followed the line of general exodus from the dining hall, and saw at the bottom of the stone steps two long troughs where groups of recruits were busy washing their trays and mugs. I came up to the troughs and my stomach heaved slightly. The water was almost viscous, barely warm, and morsels of food, fat and tea leaves floated on the surface. I dipped my tray, brought up a good deal of bacon rind from some previous meal, and went quickly to the cold tap gushing into a wooden sink beyond the trough. The fat congealed on the tray and on my knife and fork. The mug was easier to clean at the ice-cold tap. To dry them, I adopted the common pattern of swinging each vigorously to and fro. When I returned to the company lines I went surreptitiously to the ablutions, rinsed again and then rubbed hard with pieces of coarse brown toilet paper. The tray still had a fine coat of impervious grease.

Behind the camp I had noticed a steep, tree-clad hill, with conifers dotted about among the bare sycamore, oak and chestnut trees. It lay just beyond the soccer pitch, and I wondered whether it was forbidden to take a walk there. I could either ask at the company office or simply go, without inquiring. The first course seemed to be the wisest, and I went along to ask. All the offices were locked, the place deserted, but peace was again despatched by the orderly corporal strutting between the billets. Catching sight of me he called out: 'You there! I've got a job for you.' I went up to him quickly.

'Ablutions,' he said. 'I've got you on ablutions. Shit-houses included. Draw one bucket from company stores, fourteen hundred hours sharp. One brush, disinfectant, soap. Know where the ablutions are? Right. I'll inspect at fifteen hundred, sharp. Clean as a whistle, or you're on a charge.'

I was beginning to dislike the orderly corporal. There was a shrillness to his voice which grated. But I carried out his orders. The CQMS swivelled the limp stores book about once again as I signed for one brush, one bucket, one bar soap, carbolic, one canister of disinfectant.

I swilled and scrubbed around and beneath the cracked hand basins. The toilets were ranged along the end wall of the ablutions, each composed of a narrow cell divided by partitions which ended about a foot from the ground. I picked up some of the heaps of toilet paper which had cascaded from the small wooden boxes behind the doors, and threw out the sodden pieces. Then

I scrubbed again. One toilet had been blocked, possibly for several days. The smell was noxious. But I remembered the inspection due in fifteen minutes and finally went in, pouring buckets of water into the closet, holding my breath as I did so, avoiding too close a look in the meantime. The job became a succession of sluicing operations, followed by quick exits as I burst out to let my breath explode for a few seconds. In the final five minutes I peered at one or two of the graffiti on the wooden partitions. The obscenities were directed at particular sergeants or corporals. They were brief and to the point.

Shortly after three the orderly corporal banged his way into the ablutions and I stood with my brush at my side, held somewhat awkwardly, like a fusil at the 'order arms'. He poked beneath the wash basins, ran his finger around several of the encrusted bowls, flushed a couple of toilets. To my surprise, he seemed satisfied. But to have admitted this might be to sacrifice some tincture of authority, so he added, 'Next time draw some Vim and get those wash basins bright.' He probably knew as well as I did that the crustaceous layers on the bowls were now cemented to the china, and that nothing would bring them off. But he left it at that.

'Right,' he added. 'Draw brushes, sweeping, and have a go at your billet. End to end. Never let anybody see you idle, or you'll get a bollocking.' I put two and two together to make sense of the neologism and nodded vigorously. 'Yes, corporal,' I said. He stamped off, the scarlet tassel of his sash bouncing on his backside.

I brushed the billet with slow, deliberate strokes and opened the windows to let the rising dust escape. From the barrack square came the unmistakable sound of the regimental sergeant major drilling a large gathering. In the dining hall I had heard some recruits talking of their 'passing out' parade on the Friday, and I assumed they were now being brought up to scratch by the RSM. Barrack sounds seemed to be sharp, angular, piercing— whether they were shouted commands, the frequent noise of boots on gravel, or the noise of buglers practising in a shed at the other end of the camp. I wondered whether there was any music in the camp apart from military marches to the brass bands.

No further jobs had been set for me, and by four o'clock or so I suspected it might be better for everyone, myself included, if I made myself scarce. Already I concluded that the army liked you to be visibly occupied or else out of sight and out of mind. I knew better than to read my book again. The act of reading had clearly irritated the corporal. I knew now that during parade hours reading was not considered a gainful or improving occupation.

I walked over to the NAAFI and found it closed, but a sprinkling of notices at the entrance indicated the sparse attractions of the camp. There was something called 'Sandes Home', and an education centre with a library and a quiet room. This sounded more hopeful, so I studied the plan carefully.

The Education Centre was open but deserted. A few pamphlets stood on trestle tables. A pile of elementary German language texts lay on top of a bookcase. The door marked 'Library' was secured by a hasp and a lock. I looked at some copies of *Soldier* magazine, then at a map of the world on one wall. A sergeant with spectacles and blue shoulder flashes reading 'RAEC' appeared. I asked whether it was possible to browse in the library. He regretted that it was closed during parade hours. Everything in the barracks seemed to be subordinated to 'parade hours'. When did parade hours end? Officially at eighteen hundred hours. With reveille at six a.m., this made it a twelve-hour day. But the sergeant—who had none of the aggressiveness of the orderly corporal—mentioned that the library and reading rooms in 'Sandes Home' were open all day. 'You could disappear there,' he said, with a welcome flash of understanding. At last I had found someone on my side.

Sandes Home turned out to be a friendly, rambling structure, painted green, with several wings and gables. Inside there was chintz and shiny linoleum. The windows were large and pleasant. There was a reading room and a cafeteria, where middle-aged ladies were drawing cups of tea from an urn and selling scones. I decided instantly that this would become my sanctuary in the weeks ahead. The only difficulty (I discovered later) was that during the evening it became the sanctuary of almost every recruit off duty. I subsided into an armchair and found battered copies of the *Sphere* and *Illustrated London News* to occupy the next hour.

The library at the Education Centre was a disappointment when it opened at six. A few biographies of statesmen, mostly empire builders and high Tories, fitted oddly between rows of detective stories and books of instruction in languages and in practical subjects.

The evening meal, called 'supper', at the cook-house was un-appetising. Clearly the midday meal was the main meal of the day, supper only a token affair. The servers offered a hot, 'all-in' soup, tipped into your pint mug. Huge lumps of mouse-trap cheese lay on a tray, with thick slabs of bread. There was no sign of butter or margarine. Following this, and according to whether you chose to wash your mug (though some did not bother), an urn of lukewarm cocoa stood on a table in the mess hall, where you helped yourself. The cocoa was weak, tepid, barely sweetened. Probably it needed to be stirred in the urn, but there was no sign of a ladle. Another sight which became familiar at suppertime were the sparrows hopping about the many empty tables, picking at the half-eaten slices of bread. They were bold and hopped on to adjacent chairs, turning their heads this way and that as you picked at a slab of bread or sipped the watery gruel.

The troughs outside were worse than at lunchtime. The water was now cold. Fortunately, it was dark and I could hardly see the details of the floating waste. But I seriously wondered whether I would be able to stomach this daily ritual; or whether there was any machinery of complaint. Apparently there was, but im-provements were always purely temporary. The cooks hated the job of cleaning out the troughs, and blamed the recruits for not emptying their plates into the swill bins close to the cook-house wall. As the swill bins themselves were invariably overflowing and odious, there was no inclination to clean trays thoroughly. The situation was an impossible one, and most had come to accept it as one of the hazards of life at Palace Barracks.

I was still hungry and decided to seek something at the NAAFI.

There was a long queue, which explained the relative emptiness of the main mess hall. The atmosphere here was warm and cheerful. NAAFI girls in blue smocks were handing dishes over the hot-plate, mostly sausages and beans, steaming hot. The more attractive girls dealt neatly with the badinage across the hot-plate.

I finished my beans on toast and tasted NAAFI tea for the first time—I can still recall the distinct, inimitable taste—then returned in the cold darkness to the billet. The naked electric light bulbs—one, two, three, to a billet, in a straight line down the centre—glared starkly in the eyes. The radiators were stone cold and the temperature was dropping sharply. I juggled with various radiator valves, tried to trace the system back to the ablutions and beyond, but lost track in the blackness of unlit corridors connecting the six sections of the spider. Back in the icy billet I took out the stiff-backed notebook I had bought to record military experiences. I walked down to Sandes Home once more. The night was starry and a frost was developing fast. The hill above the camp was silhouetted beneath a crescent moon. The daytime noises of the camp had ceased.

I wrote up my journal in a deep chair in the reading room, anxious not to be seen, huddled over the pages whenever footsteps approached. I had no wish to invite inquiries, as I was unsure of the impression it would make. But I took my time and set down the new, at times rather frightening experiences of my first day in the army. When I had finished writing, the thought struck me again that I had made a terrible mistake in volunteering for the army. Back in the billet I slept wearing my shirt and a pullover beneath the pyjamas I had brought, and with two pairs of socks on my feet. The night was bitter. I slept fitfully and was glad when reveille sounded at six in the pitch darkness of the winter morning.

C

CHAPTER 3

Into the Crucible

A lance-corporal from the company office took me down to the main quartermaster's stores to be kitted out shortly after breakfast. A long counter stretched before us and I drew the stiff white kit bag along it, thrusting in the various pieces of clothing, webbing, and more mystifying objects the sergeant passed over. He read aloud from a list as he did so. 'Drawers, cellular, other ranks, two pairs; blouses, battledress, khaki, other ranks, one. Towels, white, one.' The sergeant handed over an object which he called a 'hussif'—a small cotton pouch containing a thimble, a packet of needles, a reel of cotton and a supply of grey wool. 'One housewife,' the lance-corporal enunciated as the object was handed over and recorded with a tick. The two boot brushes seemed to be excellent quality. They were, and I was to discover that, whatever else it does, the British army takes care of its soldiers' feet with an almost fetishist attention. Perhaps this was a tradition learned from dusty marches in deserts and jungles abroad. Certainly the longest part of the ritual of kit collection was occupied with finding two pairs of comfortable boots. It was novel, almost intimidating to have a sergeant and a lance-corporal bend down, pressing their thumbs across and around my insteps as I stood up in the stiff black boots.

'You're sure they're comfortable now?' they repeated at intervals. Other pairs were brought, until the floor was littered like the entrance to a mosque. My ankles and insteps were pressed and kneaded with intense concern.

'Well if you're quite sure, then. Only yourself to blame, mind. There's nothing worse...' I stamped around some more, and we finally agreed on two pairs, one to be 'best boots', the other 'second-best'. It was up to me to decide. The corporal offered another word of advice. 'Make these your best,' he said, rubbing his thumb across a toe cap. 'That'll bull nicely. Use a toothbrush and bone them.' I nodded, ignorant of his meaning, but I knew now that one could pick up information by watching others as much as by asking. The army seemed to prefer it that way. It saved conversation, and contributed to the general economy of language. Language was reserved for words of command. Casual conversations were at a premium.

The kit bag was now full of strange treasure. The sergeant placed a steel helmet on my head, patted it heavily, tried another and, when it ceased to wobble, announced, 'That's it. That's the lot. Sign here.' He pushed forward the clip board with its long list of ticks in one column and I signed at the bottom. 'If you lose anything, you pay for it,' he added without sentiment. He was simply recording another fact.

On the way back to the billet the corporal said, 'You'd best get the marking kit straight away. Company stores. Mark all your kit with your last three numbers. *Every* little bit. Boot brushes and all. If you don't, they'll get nicked.'

The CSM had obviously detailed the lance-corporal to help me and by now we had struck up some sort of relationship. He was a regular soldier awaiting a posting to his regiment, the Inniskilling Fusiliers, who were then abroad. He was fairly young, and had completed his training only a few months earlier. Though reserved and taciturn, he was friendly enough in a spasmodic, occasional fashion. At the billet I asked for his help in sorting out the contents of the bulging kit bag, draped like a vast white sausage over my shoulder.

We laid it all out on the bed. I could recognise the large pack, the small pack, the gas mask, pouches, water bottle, and various items of clothing, from the battledress to harsh khaki shirts and white cellular drawers. But I was unequal to assembling the various brass buckles which transformed a mass of webbing into some sort of battle order. The corporal explained and expounded, his patience mixed with condescension as he assembled it piece by piece. The small pack and ammunition pouches seemed uncomplicated, but the method of attaching the large pack on the shoulders, with a small pack at the side, the webbing straps going in and out of pouches, water-bottle container, and bayonet scabbard, baffled me. Getting it on to my shoulders involved Laocoon struggles and contortions. Getting it off involved a Houdini act.

After various gymnastics, with brief moments of unspoken panic on my part, I began to see some order in the maze of straps. Then the corporal showed me how to drape the whole contraption over some nails sticking from the wall behind my bed. The steel helmet crowned all. He left me to mark my kit and this occupied the rest of the morning. I was unmolested. At first the sticky black marking ink and the crude marking sticks left a smudgy cuneiform on the back of webbing straps, belts, and gaiters. Things improved as I went on. Finally the last three digits of my army number were inscribed on about three dozen items, including drawers cellular, braces, and the cotton flap of my 'hussif', though here the ink was promptly absorbed and the digits illegible.

When this was done and my kit laid out to dry on the floor, I went to company office to collect my pay book, AB64 Parts I and II. One part seemed designed to remind me of my oath of allegiance, with some provision for making my will on active service. The other was for recording my weekly pay. This was twenty-eight shillings a week. Fortunately I did not smoke and had never drunk beer or liquor. In the weeks that followed I began to taste beer. Resistance would have been construed as unmilitary prudery. But I avoided tobacco.

The lance-corporal told me to blanco my kit as soon as the marking was dry. When I again showed my ignorance of the process he looked astonished. 'I'll show you first parade after dinner. Fourteen hundred hours,' he said. 'Buy a cake of Number Three, Green, from NAAFI stores this dinnertime. And a brush. They'll tell you what sort.'

The rest of the afternoon was spent applying three or four separate coatings of blanco to all my webbing. The corporal showed me how. I sprinkled the water liberally, as he directed, then scraped away at the cake of blanco and brushed the wet,

green chalk vigorously into the outside surface of the webbing.
Initially I was about to apply it to both sides, but he stopped me
and directed otherwise. 'Not the inside. You'll get it all over you.
Give it several coats till you build up a surface. Like this,' he
pointed down to his webbing belt. 'And smooth. Not patchy.
Don't use too much. A bit at a time. Build it up slowly. Belt and
gaiters are the important things. You'll be inspected every day.
Concentrate on them.'

Other scraps of useful advice were thrown in as I brushed. He
probably enjoyed the instructional side of it all. He wanted to
rise to Warrant Officer First Class, he told me at one point, but
I had no idea what this signified. I became more and more grateful as new mysteries were unlocked. In the billet at the end of
the afternoon he showed me how I should polish my boots and
the leather straps of my webbing gaiters. With his index finger
inside the yellow duster I had bought at the NAAFI stores, he
took up the polish, touching it very lightly; then he spat on the
toe cap of my boot and rubbed gently with a circular motion.
Then he buffed it with a dry portion of the soft lint. I was deeply
impressed with the result. He watched my attempts, now and
then seizing my finger and saying, 'Gently! Gently! You're rubbing too hard. Like this . . .' and he demonstrated once more.

'Remember, you'd better bone your best boots. Plenty of time.
You won't be wearing them for a couple of weeks. Get an old
toothbrush. Use the handle. Warm the polish. Then rub it on
hard. Bone like hell. Get rid of these nobs,' he said, fingering
the grain of the leather. 'Bull' was a means of passing the hours
in barracks. It meant you were never idle.

I was now expected to wear a beret, webbing belt, boots and
gaiters. The lance-corporal showed me how to pierce the beret
to fix the badge of the Twenty-Eighth Training Battalion, then
he adjusted the beret on my head, this way and that.

'Hat-band straight across the head. Horizontal. Dead straight,
or the RSM will give you hell. Cap badge one inch above the left
eye. That's right. Pull the beret down a bit at the back. It'll
shrink a bit when it rains. That's expected. It'll look better. It's

too flipping floppy right now. (He used the common argot of the barracks.) 'That's better. You're beginning to look like a soldier.' At that he left.

I took up my boots and found them snug, though stiff as yet. The gaiters, newly blancoed, were also stiff. Then an awkward problem arrived. I could not remember whether the gaiter straps faced the front or the rear. It could be either. It seemed easier to pull them forward through the buckles and tighten them about the ankles. I tried this, but although they looked satisfactory something about the shape round the ankles suggested that each gaiter belonged on the other leg. I discovered they were upside down. I turned them about, and found greater comfort. This settled the matter, though I took care to check with a glimpse of some recruits as they passed the window.

For most of the following six days I was on 'fatigues' of one sort or another. One morning it was collecting bread from the bakery and distributing it to the two main cook-houses and to the sergeants' mess. The driver sat in his cab as I flung dozens of loaves into the back of the lorry where, so far as I could see, they mingled with coal dust and sand. Another day I joined a small squad detailed to carry out 'coal fatigues'. This involved heaving coal from an open yard into four or five waiting vehicles. It was heavy work and we were equipped with enormous shovels. The corporal in charge was another tyrant, who stood close by to see that our shovelfuls of coal satisfied his sadistic standards. As we worked and sweated, there was a complete absence of the usual banter which could lighten fatigues. We heaved in silence, our faces and hands gradually blackened, our eyes standing out like music hall minstrels'. The coal dust permeated all our clothing. When we finished we were allowed a visit to the bath house. But despite a succession of baths the coal dust chafed until the regulation change of denims was allowed at the end of the week.

By then I had gathered a good deal of information about training in the weeks which lay ahead. Fatigues encouraged comradely exchanges among those I joined for the day's work. They were mostly recruits half way through their basic training and had been put on fatigues for minor acts of indiscipline or idleness. Putting their accounts together, I learned that basic training would occupy twelve weeks—a good deal longer than normal basic training elsewhere; but then, they said, the Twenty-Eighth was meant to be special, designed to train future NCOs and officers. Apart from being much tougher—unsparingly—you learned how to give orders, how to take a squad at weapon training or across country, and even at drill on the square. You were not allowed out of barracks until you had 'passed off the square' —usually in the fourth week—because the commandant did not allow anyone to wear the king's uniform in Belfast unless the recruit could carry himself like a soldier, meet standards of turn-out suited to the army's future leaders, and knew how to salute officers.

Other pieces of information came my way in the NAAFI and in the mess hall, so that by the end of the week there were compensations for my mistaken zeal in reporting a week early. At least my kit was marked: the galaxy of brass fittings and buttons spread about the webbing and packs and pouches now glistened. My best boots had achieved a dull, burnished effect—but not the mirror-bright brilliance of some of the NCOs who marched about the camp.

Immediate authority and hierarchy were represented by warrant officers and NCOs. The world of the officers seemed at a distant, olympian remove. I had the impression that all the work, certainly all the discipline, was carried out and maintained by sergeants and corporals. Officers occasionally appeared, but only in order to return salutes by tapping their peaked hats with leather-clad sticks. Their real function escaped me; they seemed some sort of adornment, to give the place tone.

Once or twice I saw the company commander around Baker Company lines, a slightly rotund major with a sandy moustache, pale blue eyes and an air of having been to a minor public school. He looked vaguely worried much of the time, though later I heard he had a distinguished war record, like many of the other officers at Palace Barracks. A more striking character around Baker Company was a very tall paratrooper, a lieutenant who carried a blackthorn walking stick and wore brown boots. His moustache was considerably bigger than the major's. He was popular with the permanent staff and his general swashbuckling air made him something of a favourite as a platoon commander. He had a bellowing laugh and was very profane.

By the end of the week, therefore, I was thankful to have had the chance to ease myself into a strange, at times brutal environment. Once or twice I caught sight of the white-belted, white-gaitered military police marching small squads of men at a frightening pace. You did not have to be told that they were on 'Jankers' for some misdemeanour, or possibly for mere idleness.

I was also glad that one unpleasant incident was now past me. On a bitter morning, as I went for breakfast in the pre-dawn darkness, I again caught sight of the two troughs outside the mess hall under the stark electric light. Some of the recruits were already dipping their aluminium trays and swishing them about. A good deal of bacon rind and white of egg eddied about the black surface of the tepid water, and the nauseous smell caught me as I went up the three stone steps into the mess hall. In the fitful light inside, the warm cloying smell of grease seemed even more rancid and penetrating. At the hot-plate the servers slapped the hard fried egg on to the cold aluminium. The bacon was thick, underdone.

As I went to a table with my brimming tea mug I could see that everything on my tray had congealed; watery porridge was lapping around the bacon and egg. The thick slice of bread stuck in my throat as I tried to stomach it, though gulps of sweet tea helped it all down. Once or twice my stomach heaved, and I retched slightly. Others about me were eating unconcernedly. Obviously they felt an urgent need to stoke up for the rigours of training on a bitter day. As I got up to leave, the retching continued, though I just managed to contain it. Then, as I approached the open doors leading down to the zinc troughs, first the bare electric lights above and then the whole mess hall pivoted. The world went muzzy and the open doors before me zoomed to right and left. Somewhere I thought I heard a bang and a tinkle.

When I came round they had helped me to a sitting position in the cold air at the bottom of the stone steps, my head resting on my knees. I felt very ill and queasy. Someone was piling the broken pieces of my pint mug neatly by my side on the stone steps, as if for some funeral rite. Others passing in the half light asked, 'Is he all right?' I nodded miserably without saying anything, but managed to thank the two or three who crouched near me. One of them accompanied me back to the billet, where I lay on my bed.

'You were lucky,' my helper told me. 'You might have gone head first down those steps. One of the blokes saw you were going. You'll need a new mug.' I thanked him again and told him I would be all right now. It had passed, and all I needed was a bit of a rest. 'I'll tell them at your company office,' he offered, and left, closing the billet door quietly. I was grateful for the fellow feeling, my first real encounter.

After half an hour or so I was feeling slightly better. Cold air came through the open window and my brow seemed much cooler. The CSM appeared, and assumed a bedside manner. I decided that the army was not as inhuman as most of the week's work had suggested.

'Do you want to report sick?' the CSM asked. I did not know what this involved, and asked him rather weakly.

'Not much, but if you can't parade, it's regulations. Down to the orderly corporal's bunk: battledress, plimsolls, small pack with your wash kit. You'd better rest for a bit, then decide.

'Thank you, sar'n major, I'm sure I'll be all right by then.'

'Don't worry about it. You'll need a new mug. You'll have to pay for it,' he said, eyeing the shards whimsically.

The CSM gave orders that I could rest until I felt better. I dozed until ten and then made my way to Sandes Home, for coffee and home-made scones. By lunchtime I had recovered, but I took my meal at the NAAFI. Next day I was able to face the mess hall once more.

CHAPTER 4

No Nonsense

The new intake arrived on the Monday morning. National-servicemen made up the bulk of each intake but 'regulars' were also included. What was common, it seemed, was a 'potential officers' status conferred by educational level or by the recognition of some leadership potential, so we were told. Intakes arrived in groups, probably staggered by posting orders. They made a curious spectacle in their plain clothes, roughly assembled in threes, and carrying a great variety of suitcases, with a sergeant in attendance telling them to keep in step. Shuffling along, they looked like displaced persons, or refugees in flight.

Outside the company office the sergeant drew them up and called out 'Right, turn!' which produced an untidy rippling effect among their ranks, and a few maverick turns to the left. Earlier arrivals were chivvied out of the billets. I was told to join the first group, and slipped myself at the end as unobtrusively as possible. The drill sergeant walked along the front rank, his hands behind his back, spelling out orders for the day. Two more platoon sergeants hovered about, and a number of corporals and lance-corporals assembled, most of whom I had seen stepping about the company area in the previous few days. Then the CSM appeared and ran his eye over the throng. Some were tall and thin; some squat and tough looking. I noticed the gentle faces and elegantly tailored clothes of some; visible poverty and rugged, weatherbeaten faces among others. Potential leaders sent to the Twenty-Eighth Training Battalion were clearly drawn from a broad cross-section of British society, though I had the impression that National-Service public schoolboys in well-cut clothes predominated in the fifty or sixty recruits drawn up.

One of the sergeants produced a list and told us to answer our

names. Where the name was a common one—a Jones or a Smith
—he added the recruit's last three numbers. Our initials were
omitted in the rapid series of barked names and answering 'Sirs'.
We were told to learn our army number, to be prepared to rattle
it out whenever an NCO or an officer asked it. This was one of
the ways in which the army taught new recruits to be alert, as
well as to react to an order. In the next few days, the command
'Name and number?' was flung at individuals at all times of the
day, usually for information, sometimes to test alertness or any
tendency to panic. Some stumbled over their allotted number.
This provided a corporal or a sergeant with a useful opportunity
to mix incredulity with scorn and mild abuse of the recruit. The
expletive occurred in every sentence, as adjective, noun or gerund,
producing bizarre syntax here and there. 'Don't know your own
fucking number? What fucking next? Fucking forget your own
fucking NAME next.' I was now familiar with the tendency of
NCOs to stress words at intervals, without any apparent verbal
consistency or logic. But it gave sharpness to commands and indeed to any sort of comment. It could also achieve marked effects
as the days went by and sergeants began to discover our separate
weaknesses on parade. By then we were a step nearer to one of
the peculiar veins of army humour—the singling out of the nervous, the weak or the stupid to serve as a butt for the sergeant's
stock of insults and barbs.

When the roll call was taken on this first day of the new intake,
the sergeant detailed us into platoons. When these were allocated
he read out the name of the platoon commander, the platoon
sergeant and NCOs of each platoon in turn. We were to get
these names inside our brains immediately, said the sergeant, or
life would be that much more hellish. The platoon corporals led
us to the empty billets. The newcomers were told to drop their
suitcases in the billet and report to company stores for bed and
bedding. The atmosphere was one of rapid reactions and movement. We moved.

By mid-morning our billet contained about twenty beds,
approximately in line, and our platoon corporal was demonstrating the art of boxing blankets. The sergeant looked in now and

then, cocking an eye down the line of beds, kicking one or two
bedsteads which offended his eye. He was in the Devon Regiment
and had a sense of humour. All NCOs and warrant officers seemed
to be divided into 'bastards' and 'ok types', according to mess hall
lore. Sergeant Channing was in the latter category. He also had
a west country accent which contrasted agreeably with the Irish-
ness of most members of the permanent staff.

The lance-corporal put in charge of our billet had finished basic
training some weeks earlier. He was a tall, gangling, extremely
youthful looking public-school man awaiting a call to the War
Office Selection Board. He seemed nervous, and over-compensated
for this when giving orders. He was quickly christened 'Baby Face'
by two Glaswegian recruits, though not in his presence.

By noon the whole complement of the intake was at company
strength and most were struggling back from the QM stores
weighed down by bulging white kit bags. I was more than ever
glad to have got this over, and the sight of my kit assembled on
the wall, brasses polished, brought one or two expressions of envy,
besides making me a minor counsellor when the lance-corporal
was not demonstrating kit layout in his imperious, angular manner. I helped one or two to adjust buckles or to identify bayonet
scabbards and water bottle covers. By then we were chatting
among ourselves and learning of our varied routes to the Twenty-
Eighth Training Battalion.

By mid-afternoon, when everyone was busy marking their kit
and I was working away with black polish and a yellow duster
at my boots, I discovered that two immediate neighbours across
the bedspace had been at Eton together. They were going up to
Oxford after they had finished National Service. Both wished to
join the King's Royal Rifle Corps after basic training. They were
soft-spoken, restrained and utterly courteous. Beyond them was a
Scot, and beyond him a cockney, both 'regulars'. The Scot was
fairly taciturn, absorbed in his tasks; the cockney a bit of a wag,
with a stream of self deprecating comments on his own inexpert-
ness as he wrestled with buckles and buttons.

During the next couple of days I discovered that we were even
more of a mixed crew than first appearances had suggested. Eton-

ians, Wykehamists, a man from the Gorbals who preferred giving orders to taking them, a quiet, reserved and slow-speaking west countryman, and a farmer's son were among us. The most arresting figure was a tough-looking man, older than most of us, who squared up to our lance-corporal on the first afternoon. His name was Joe, as he instructed all of us to call him. He had been in the navy before enlisting in the army. I wondered who or what had directed him to Palace Barracks. Possibly some army selector had noted leadership qualities lurking behind the aggressive manner. Perhaps he would make a good weapons instructor eventually. When we were issued with rifles he seemed highly competent in stripping the bolt mechanism.

So we were an odd mixture sprinkled about the billet. One man prayed by his bedside for a long time each night at lights out. Another could not be awakened in the morning until the corporal took hold of his bed and turned out the snoring figure, bedding and all. A certain type of humour was struck up among the group, mostly to do with our common predicament in being universally considered the lowest of the low throughout the camp —that is to say, raw, untrained recruits who could not march correctly, nor salute, nor look like soldiers in our denims and floppy new berets.

At this stage there was no quarrelling, nothing abrasive in our relations with each other. The first troubles arose as the gruelling programme of parades, weapon drill, map-reading instruction, and more and more drill began to fray our nerves. Some had already received their first 'bollocking' on parade, for sloth or a dirty turn-out. Every evening a good hour or so was needed to blanco webbing, shine boots, cap badges, and brass buckles on belts and gaiters. One day, someone was missing his tin of Brasso. He bought another after cursing mildly. But a frisson went about the beds when Joe, our navy veteran, announced that he couldn't find the duster he had bought at the NAAFI shop. Joe's bed and locker were right next to the door. He had arrived later than the others on the first day. The old injunction about sleeping near the door came to mind. But whereas others had toured each bed after 'losing' an item, inquiring with some degree of courtesy,

Joe's method was rather different. 'Which bastard's got my bulling cloth?' he demanded loudly. Several looked up from where they sat and shook their heads. One or two close to Joe gave negative replies.

'I want the bastard who's nicked my bulling cloth,' he shouted, his hands on his hips.

There was a stir at the far end of the billet, where the lance-corporal slept behind the two tall green lockers which partitioned his bedspace from the rest of us. Now he appeared, rather incongruously attired in a dressing gown and slippers.

'Stop that bloody noise,' the lance-corporal said. Joe pivoted. It became a silent trial of strength. Joe gave out to a good deal of bad language about thieves in or out of the billet. The corporal said, 'Come here!', and as Joe went up silence fell in the rest of the billet. The corporal had the advantage of height, though at that moment he was hardly dressed with authority. But he knew that this was now a public trial of strength.

'Any more of that, and you'll be on company orders tomorrow.'

'What charge?' Joe snarled. He knew his rights, even in the military.

'Refusal to obey orders. And insolence, if you don't watch it. Now get.' And the corporal's fingers pointed up, his arm following with a quick stab at the air. Seconds passed, then Joe turned round quietly and went back to his bed. He muttered among those nearest to him, but kept the tone low. I felt this would not be his last encounter with the lance-corporal, and it was not.

Baker Company drilled on a smaller square, safely quarantined from the RSM's private acre, where only the elect could appear after several weeks of daily foot drill. It is difficult to recall our first fumbling attempts. Memory has somehow erased them. Clearly, we must have been appallingly bad, although some had been to OTC at school and could recognise the commands. Sergeant Channing took us every morning of the week at 0800 hours, including Saturdays, and there was usually another hour in the afternoon.

I did better than I had expected or feared. I was very fit and could think quickly, so I had sharper reactions and reflexes than some. I learned to anticipate, or otherwise watch the heels of the OTC men and adopt their reactions within a split second. We were sized from left to right, and although there were early attempts to avoid the front rank by scrambling for the rear, these were scotched by the drill sergeant's trick of going to the rear, shouting to our backs, 'About TURN! OK. You lot are now the *front* rank. Pick up your dressing...'

So the jockeying came to be concentrated on the middle rank, which seemed safest in the face of this pincer threat. But here again the sergeant knew the old tricks and would watch out for faces which always contrived to begin the day's drill in the middle of the centre rank. They would be winkled out to the front rank with Sergeant Channing's combination of sternness and good humour. 'How come you're always in the centre rank, Jones 112? Shy, are you, Jones? I'll have it in for you if you're not careful. Move forward. Front rank, here. Drop back, you. Saw enough of *you* yesterday.'

A good deal of banter went on, and, after a few days, various obscenities were introduced. Eventually, one realised that one of the few rewards of the drill sergeant's life is to test the spirit, will, and perhaps the sense of humour of his flock on the drill square by a stock of obscenities fashioned and sharpened over the years. They mostly referred directly or obliquely to sexual assault. Vivid euphemisms for sexual organs, male and female, appeared and reappeared, often with impressive invention and eloquence. The whole canon was a tribute to the richness of the English language.

Every platoon, every section had its scapegoat, and you hoped it would not be you. Our scapegoat was one of the two Etonians, who, despite his charm and good manners, was almost pathologically untidy in his dress and general appearance. Worse, he did not seem to be entirely in charge of his own limbs, which let him down badly as the drill became more intense and the pace quicker.

Drill began with Sergeant Channing's inspection of the three ranks as we stood to attention. This provided a suitable overture for the flow of expletives as we drilled later.

'Fortescue, take your arse in! Mustn't stick it out like that
man. It's dangerous round here. There's old soldiers hanging
about. Might take advantage. Might fancy you myself, you go
on like that. You're my type, Fortescue. Keep it in, then. No,
DON'T stick your belly out. Like a pregnant nun now. Send
you to hospital. They'll operate. See what sex you are. Bit of a
mixture, Fortescue, that's what you are. What are you?'

'Bit of a mixture, sar'nt.'

We all knew the drill for these rhetorical questions. No one
suffered unduly, though some came in for more insults than
others. Yet this was one way of goading them to do better. You
were not allowed to show any unusual or eccentric characteristics,
good or bad. Individuality was banished, if only for the duration
of the drill hour. Overall, it was probably good for the character.
Arrogance and idleness were rapidly despatched and we all felt
much better at the end of the drill.

When we were issued with our rifles a new crop of epithets came
from Sergeant Channing's verbal bank. As we struggled to shoulder arms, and generally to manage our rifles like weightless toys,
as we were told we should, Sergeant Channing would look dis-
trustfully at us:

'Some of you lot look weak. Self-abuse, that's what it is. Too
much wanking under the blankets. Leave it alone, you should.
Go to bed with boxing gloves if you can't leave off. You'll get
wankers' doom. Horrible fate. Too horrible to mention. Know
what wankers' doom is, Jones 112?'

'No sir.'

'Well, you're well on the way, Jones. All these public-school
capers. That's what's up with some o' you. And I won't have
sodomy in the ranks, so keep your distance when you're marching.'

It was all fairly good-natured. Sergeant Channing had drilled
many hundreds of public-school men, and he knew the lore. The
disparities of home or school background had hardly any effect
inside the billets. We were all too much occupied with the stark
problems of survival. Within two weeks most of us were swearing

cheerfully in the familiar argot. One quiet man resisted for a few more weeks, but the need for quick, direct communication in a common language between everyone brought his capitulation in turn.

We were marched everywhere: to meals; back from meals; to company lines; to weapon drill; to the gymnasium; to bayonet practice. After three weeks we began to show some sort of coherence, though not as much as those who were weeks ahead of us in basic training. The .303 rifle also began to seem lighter. We could slope arms, shoulder arms, present, even port arms. The parade-ground banter continued. 'Watch the way you push the barrel out when you come to the high port. *Don't* injure the man in front of you. No sodomy in the ranks. Army regulations...'

On weapon drill, we learned how to crawl on our bellies with the rifle between the thighs. It was a dangerous position. Once or twice the rifle bolt caught the scrotum, and on icy January mornings the effect could be devastating. Sergeant Channing was ready with the correct homily when a gasp of pain came from the frozen ground. 'Watch it! watch it! Don't lose your goolies, man. Can't have eunuchs in the British army. Bigger the balls, the better the man. Regimental motto, that. Which regiment? He swung round on the rest of us suddenly and pointed at me. 'Which regiment, *you*?'

'I don't know, sar'nt.'

'Beds and Herts,' he said, without a smile. No reaction. He paused. 'Use your loaf, Beds and Hurts!' he shouted. Then we saw the awful pun and laughed dutifully.

After we had crawled with our rifles beneath cruel hedges and bramble bushes beyond the soccer pitch, we graduated to other light weapons in the infantry armoury: the two-inch mortar, the Sten gun, the Bren light machine gun; and finally the PIAT— or Projector, Infantry, Anti-Tank. The PIAT, later superseded by the 'bazooka' rocket launcher, was a fearsome weapon. It was extremely heavy and cumbersome, with a vicious kick as some vast spring inside shot the heavy iron spigot forward to launch the anti-tank bomb. Smaller members of the squad sometimes finished

D

at the bottom of the firing trench, flung about by the iron monster. Some closed their eyes as they pressed the trigger, but the
corporal instructing us was familiar with the habit. 'What's the
use of that?' he demanded. 'You'll get a bayonet in your balls,
sharpish, if you go round like that.'

There were ribald moments with the Bren machine gun. When
carrying out 'rapid fire', which involved changing the magazine
at high speed, the eyes had to be kept on the target throughout.
But most of us tended to look down at the gun to switch magazines, then look for the trigger again. Sergeant Channing dealt
with this offence with a crude idiom: 'DON'T look DOWN!
FEEL for the trigger. I'll put some hair round it, then you'll feel
it alright, *I'll* bet. Use your walking-out finger!'

The term was later explained to the innocents by sexually experienced men like Joe. The military education progressed. The
brutalising process was furthered during bayonet practice. This
involved charging, with bayonet fixed, at a series of sand-bags
hanging from miniature gallows. As we lolloped forward, one or
two missed the swaying bags completely, with comical results as
they lunged forward, their bayonets biting deeply into the frozen
ground ahead. You had to recover bayonet and rifle sharply, as
the next man was already bearing down a few paces behind, ulu-
lating on the run-up, screaming at the final lunge. If he also
missed the swaying, straw-filled bag, his bayonet was likely to
come close to you from behind. Bayonet charges could even bring
out some latent murderous instincts in one or two of the recruits.
Joe's blood curdling yells outdid the corporal instructor's war-
cries. As he jabbed his bayonet into the bag, one wondered if he
had in mind our tall lance-corporal in the billet.

One could distinguish those with genuinely aggressive impulses
from those who were terrified of bayonets and whose thin cries
brought derision from the corporal when they returned, sheep-
ishly, to the start line.

'Call yourself a soldier?' the corporal would say sarcastically.
'You wouldn't scare a ten-year-old tart with that lot. What you
goin' to do if you got to make a bayonet charge on a bunch of

wogs? Do it in your pants, most of you would. Shit scared, some of you lot.'

Insults were the common coin of all the instructors. It was all part of the moulding process. As a drill sergeant later told us at Sandhurst, 'You slap them down to build them up. You've got to start from scratch. No favourites. Everyone the same. Makes your men muck in better. Builds up platoon spirit. And comradeship. That's what the army's about. It wins battles.' Then one saw it from another perspective. It wasn't *all* mindless sadism: there was a point to some of it.

Certainly we 'mucked in' together, the strong assisting the weak in a hundred different ways. Part of the toughening-up process consisted of route marches along the country roads beyond the camp. These were really endurance tests, the first one four miles, then five, then six, eventually eight. Corporals from the Army Physical Training Corps set the pace, surging ahead of us in their red and black striped jerseys and tight black slacks, like exotic wasps. Another two chivvied us from behind. These were men who spent most of their working day boxing, fencing, or performing impressive handstands in the corner of the gymnasium whilst we recruits puffed and jumped on the spot. Keeping up with them on a route march was the most strenuous thing so far, although there were other tests to come. With their bunched fists rolling across huge chests these argonauts strode along, delighted to be breathing the icy morning air, visibly enjoying it all, even when they fell back to scourge the stragglers. We were ordered to keep in step, but at such a cracking pace it produced bizarre effects on different individuals. Oddly, the tallest recruits suffered most. Their long legs could not stride out as they would have preferred, and instead they had to accept the pace of the shortest, with pathetic results. One man, well over six foot, most of it legs, might almost have stepped over the smallest member of the platoon, and his long shanks worked away like pistons until, near the end of our first march, he could be seen giving way at the knees. Back in the billet his legs gave up, and they palpitated violently as he lay on his bed.

The marches did not affect me as badly as some, as I was of

medium height, neither thin nor fat. But the first march was agonising for everyone. Acute pains darted along the calves and thighs as the first couple of miles passed at a cracking pace. Boots would suddenly feel like lead, every pace a stroke of pain. The PT corporals grinned broadly at our grimaces, their forearms pummelling the air across their chests. 'You're not FIT,' they called out gleefully. 'Keep up! DON'T slacken the pace!' And they would fall to the rear, snapping at the heels of the stragglers like sheepdogs chivvying a flock.

When we marched back into camp the PT corporals dismissed us contemptuously at company lines and sprinted off effortlessly. The platoon tottered in, to await foot inspection by Sergeant Channing. Some had genuinely suffered. One cockney recruit, a short, thin man, red at the knees and grey at the gills, gasped out, 'I'm buggered', and collapsed face down on his bed. His breathing continued audibly, so we let him be. His neighbour helped him off with his boots a few minutes later. Alarming blisters showed at his heels.

Sergeant Channing moved slowly past the bare, upturned feet which we raised for his inspection. Few of us had escaped blisters behind the ankles. One or two were told to put on their plimsolls and go off to the camp hospital with the orderly corporal. The little cockney said plaintively, 'I can't take no more o' that, sar'nt', but Sergeant Channing was unmoved. 'Can't leave here without you complete your training. The medics will mend those.' He made as if to touch the blisters, but the feet darted back into a reflex foetal position.

On the next occasion, almost all of us managed the pace: an astonishing capacity of the human frame to adjust to circumstance. By this time, our blisters had developed into layers of hard tissue. Daily square-bashing worked remarkable effects on soft skin. I almost enjoyed the final route march in the crisp air. At least it got one outside the confinement of the camp. We caught sight of passing civilians as though they belonged to another species.

At the end of our fourth week, our drill apparently reached a standard which allowed Sergeant Channing to parade us on the main square. As we were about to set out he issued stern warnings.

'I'm taking you on the main square. You all know what THAT means: the RSM. Now listen, every one of you...' We could sense that Sergeant Channing really meant it, and that his own *amour propre* was at stake. Platoon sergeants were not supposed to bring recruits to the hallowed ground until they were sure the required standards could be met. For the first time a slight nervousness showed in Sergeant Channing's voice. But we were on his side, and indeed had grown attached to him by now. Inevitably, perhaps, he had become a sort of surrogate father for every member of the platoon; part feared, part admired, part loved, part hated.

We swung our arms strenuously, straining to keep together, aiming at the crispness of footwork which only trained soldiers achieve in unison. As Sergeant Channing formed us up on the main square I caught a glimpse of the RSM cruising and brooding in the shadow of the chestnut tree beyond the asphalt. Now he was standing very still, his drill stick under his arm, unmistakably watching our every move.

We did our best. We marched, counter-marched, changed step, did several about turns, saluted to the front, saluted on the march, marked time, slow marched. Sergeant Channing kept up a degree of braying very rare for him. No doubt he was hoping to get by without intrusion from the distant figure, who was now showing distinct signs of excitation at the edge of the square as he paced back and forward, his face turned towards us all the time. Then the storm broke. An extraordinary volume of noise came across the square. It rolled back from the walls of the clock tower to the motor transport sheds, and bounced over us.

'SAR'NT CHANNING!'

'SIR!' Sergeant Channing had spun around, facing the source of the noise, slamming his feet to attention.

'SAR'NT CHANNING, GET THAT SHOWER OFF MY SQUARE. AT THE DUBBULL!'

'SIR!'

Sergeant Channing spun back again and shouted at us. 'PLATOON-SHUN! LEFFTURN! AT-THE-DUBBULL DUB-BULL-MARCH! LEFF-RIGHT-LEFF-RIGHT-LEFF-RIGHT.' We doubled off, kicking our knees high. Behind us, terrible imprecations reached our ears from the shades. 'DISGRACE TO THE BRITISH ARMY... BLOODY SHOWER... *MY SQUARE*!'

When we were well out of sight and out of earshot, a breathless Sergeant Channing slowed us down to a march. Then we were standing outside company lines, flushed, heaving, perhaps ashamed to have let him down. Sergeant Channing managed a panting smile.

'Don't let it get you down. Every other lot gets that. It's the RSM's little game.' He took us into his confidence daringly. 'If you'd been twice as good he'd've bawled you off. Gives him something to do.' We knew better than to laugh: that would bring an instant switch of allegiance in the tight code of authority and hierarchy.

But the little contretemps meant that our passing off the square, and thus any exeats to leave the camp, were delayed by several days. Sergeant Channing marched us on to the main square again a few days later. This time the RSM circled us from the rear, about fifteen yards away. Now and then he would interrupt the drill and point his stick as Sergeant Channing brought us to a quick halt.

'Sar'nt Channing, CENTRE rank, the THIRD man from the LEFT: idle gaiter STRAP.'

Sergeant Channing darted into the ranks, introducing a novel venom into his voice. 'YOU, you bloody fool Jones. Next time you'll lose your NAME. DO IT UP!' And Jones 211 bent down with twitching fingers to do up the strap.

We completed our initiation with one more interruption.

'Sar'nt Channing!'

'Sir!'

'Front rank, number four: bloody idle drill. Take his name.'

'Sir!' The sergeant stamped up to the suffering cockney, whipped a notebook from his breast pocket fairly theatrically and scribbled furiously as he shouted, 'Company orders TO-MORROW.'

At that the RSM turned on his heel regally and walked off, very erect. In the front rank the little cockney quietly muttered the universal, brief expletive. If Sergeant Channing heard him he did not show it. Somebody had to catch it, we all knew that. Your turn today: someone else's turn tomorrow.

CHAPTER 5

In the Billets

Each of us tended to make one or two special friends in the billet, sometimes from neighbouring billets. Because our bedspace adjoined, and because we found ourselves talking together a good deal as we 'bulled' boots by the hour, I formed a friendship with one of the two Etonians, Richard McWatters. Etonians at eighteen years come in different guises. Some are talkative, eclectic and win scholarships; others are snobbish, extrovert, and stupid. Others again are soft-spoken, self-effacing, bookish but not brilliant, and astonishingly honest. Dick McWatters belonged to the last category. A place at Oxford awaited him, but he was not destined to be a scholar. We came from quite different backgrounds—myself Northumbrian, educated at grammar school, independent, Celtic, voracious of books; he mild, retiring, the son and grandson of distinguished officers and officials of the Indian civil service. But we had a good deal in common. Music was one bond, poetry another.

On Sunday mornings we would walk up Grey Hill together, just behind the camp. When mid-February passed and the ground thawed he showed an astonishing knowledge of bird life. As the remaining winter visitors and the early spring arrivals flitted through the deep green of the conifers or the stark branches of oak and chestnut trees, he would name them unerringly, or recognise distant calls and the message they spelt. Up here, above the camp among the spruce and pine we tended to drop the expletives, which everyone now used down below among the billets and in the mess hall.

Training was still intense, and some could not stand up to it. At the end of January news had reached Baker Company of a recruit in another company who placed his head across the rail-

way lines down below Holywood village one frozen night. The
matter was hushed up so far as the outside world was concerned,
but the bare facts were passed around the mess hall and the
NAAFI. Word came from company offices that we were forbidden
to mention the matter to the press. Training continued. No one
had time to brood on the incident.

Other incidents during our training underlined the latent vio-
lence in our environment. One of the men in our billet, a young
farmer from the west country who detested the regimentation we
were subjected to around the clock, broke into the NAAFI one
night and stole boxes of cigarettes and a few bars of chocolate.
Somehow the military police in camp managed to trace the culprit and found the stolen goods in the box beneath his bed. As
we lined up for first parade one morning we saw the dazed
youth being handcuffed to a sergeant in the military police. This
was the same sergeant, rumour suggested, who had kicked a man
to death in the military jail at Shepton Mallet, and who had been
posted here to avoid violent retribution from other inmates. He
was a big, broad man, with the build of a professional wrestler.
His face showed no trace of the milk of human kindness. As he
took the pathetic culprit away from Baker Company lines our
section corporal asked the Red Cap, 'How long d'you think he'll
get, sergeant?' The answer was a malevolent scowl and the terse
reply, heard by each of us on parade, *'You'll* never see him again.'
A slight chill of horror went through our ranks.

The feud between Joe and the lance-corporal in charge of our
billet continued throughout our training period. Now and then
Joe would again almost square up to 'Baby Face', but he always
backed down at the last moment. He was shrewd enough to know
that the lance-corporal was backed by military law and army
regulations. Striking a superior officer was one of the most serious
offences in the book.

One day there was a particularly venomous verbal exchange
between the two. The lance-corporal ended it with the customary
threat. 'Another word out of you, and I'll put you on a charge.'
Joe retired to his bedspace once more to mutter under his breath.
When the corporal went out some minutes later there was no

mistaking Joe's threats: 'I'll fix that bastard tonight if it's the last thing I do. I'll spoil that fucking baby face of his, see if I fucking well don't.' His curses grew choicer and louder.

It was Friday, pay day, when many of the recruits tended to drink a few extra pints of beer at the NAAFI. Joe was one of them. By now we were all accustomed to his stumbling figure just after 'Lights Out' as he made the last few paces in the company lines in the dark. We knew better than to remonstrate with him, drunk or sober. No one in the billet would have cared to take him on.

This Friday night, as he stumbled to his bed half-drunk, half-sober, it was alarming to hear him still muttering the curses of that afternoon. It was even more alarming in the filtering light of the moon to notice Joe silently draw his bayonet from the scabbard on the wall above his bed and slip it beneath his pillow before resting on his bed, fully clothed. Silence came in the billet, except for the heavy breathing or snoring of those who were drunk or asleep. But some of us were very much awake: the quick nervous turning of half visible heads, even the steady breathless stillness about the bedspaces confirmed it.

Perhaps five or ten minutes passed, with Joe lying silently on his bed. It seemed much longer. Then slowly, stealthily, he could be seen getting off his bed, drawing something from under his pillow which gleamed. He tiptoed towards the end of the billet where the lance-corporal was sleeping. He made no more than three or four paces before blankets were flung off several beds and there was a concerted rush at the creeping figure. A scuffle followed, and someone switched on the light. We had to struggle very hard to wrest the bayonet off him. It vanished quickly, just as the corporal appeared around the lockers, blinking, but obviously taking in the significance of the scene almost immediately. Under the harsh electric light Joe dusted himself down and the two enemies faced each other, about three yards apart. Wisely, perhaps, the corporal sought a diplomatic retreat.

'It's after Lights Out,' he said in a level voice. 'All of you get to bed.' He ran his eyes over each of us, hardly pausing at his enemy standing between the rows of beds. Now other bedworn figures joined in a chorus of complaint. 'Get to fucking bed: noisy bastards. . .'

It was a tense moment. We were relieved when Joe once more chose retreat. Perhaps the tussle with his fellow recruits, who were, after all, his daily comrades, had some lasting effect. Whatever the case, he avoided the corporal thereafter. For his part, the corporal had clearly decided to avoid any further direct confrontation. If he did not know it precisely, he seemed to guess that, one way or another, he had had a narrow escape.

The intense pace of our training schedule increased the camaraderie of the billet. When kit inspection came round the neater, more practical members of the platoon helped an unbelievably impractical Etonian to lay out his kit on the bed. His first attempts were almost comically inept, lacking any symmetry or order. Our cockney comrade (who had kept a fruit and vegetable stall in east London, we discovered) took him in hand and showed him how to roll his spare bootlaces into neat catherine wheels in the prescribed manner. Others showed him how to press his battledress before drill parades. He was visibly grateful for these efforts. Few of us wished to remember the first kit inspection, when the company commander had first stared down at the grubby disorder on his bed, then inserted a cane under the stretched blanket and heaved the lot on to the floor before ordering two extra kit inspections for the man. It was the only time I saw tears in the billet. He sat by himself silently, trying to hide them, but the tears fell on to the kit he was polishing. Those of us who noticed weren't sure what to do. We had no wish to draw attention to it and bring him loss of face if we reacted. He rallied after a time, and next day several about him gave assistance in a deliberately casual way. Adversity made for fellow feeling.

A more sedentary part of our training period came with tests by personnel selection officers at the PSO centre in camp. These were of considerable variety, and designed, it was said, to bring out the special characteristics of each of us for different forms of leadership or command. Some of the tests were merely practical. At

separate benches, each of us was confronted by a number of unfamiliar pieces of metal and wood with nuts, bolts and screws. We
were told to assemble them into tools. The dextrous and practical-
minded quickly slipped the assorted pieces into place by finding
slots, notches, flanges and screw-threads, in some harmonious
pattern. The result could be a workmanlike tool such as a wrench,
a miniature vice, or a carpenter's plane. I was not among the
totally incapable, who struggled with their pieces, then jammed
them, or feverishly hammered them on the bench, until warned
not to by the cruising PSOs. Neither was I among the dextrous.
By patient puzzling I was usually able to assemble some sort of
tool, with a degree of malfunction in some vital part of the
mechanism.

Then there were written tests. Some of these were fairly factual,
testing general knowledge in current affairs, politics, simple economics, and aspects of society. Others tested reasoning power. We
were required to complete as many answers as possible out of a
total of three hundred or so. We were assured that any score over
two hundred was considered phenomenal, and that the average
was somewhere around ninety. I did not take note of my scores,
and soon forgot them, but I did not adopt the technique of an
Oxford philosophy don I met some years later. As a promising
scholar he was already deeply interested in philosophical problems by the time National Service claimed him. Taking his PSO
test, he noted an interesting though recondite philosophical
proposition in the first question out of the three hundred set. So
he devoted all the available time to a disquisition on question
number one. When the results were made available, he was de-
cleared educationally subnormal, and never rose above the rank
of private.

On the basis of these tests we were nominally divided into
categories. 'OR1s', that is 'Other Rank (1)s', were taken to be
potential officers. 'OR2s' were potential warrant officers. There
may have been lower gradations, but we did not learn them. I was
relieved to be placed among the OR1s, as the men I liked and
wished to continue seeing were mostly in this category. But it was
difficult to avoid the conclusion that OR1s were in the main

marked out only by superior education and written skills.

Thereafter, however, the OR1 status involved some extra hazards during training periods. Warrant officers clearly took a special pleasure in discovering that slowness, ineptness or other incapacity in practical training was the likely mark of the OR1. One drill sergeant major in the Royal Ulster Rifles, a wizened veteran who had served for more than twenty-five years, had a fine turn of phrase for the OR1 on weapon drill or on the parade ground. And he had a sort of sixth sense for spotting OR1s. His opening question was generally placid, merely inquisitive:

'Would you be one of these OR1s, now?' When the recruit nodded, a vigorous monologue followed, the sergeant major wide eyed and incredulous.

'You mean *you're* supphosed to be fucking arfissur material? You're not arfissur material, mahn, you're a dozy fucking CROW, that's phwat you are. Phwat ARE you, now?'

I'm a dozy fucking crow, sar'n major.'

'That's RIGHT. That's phwat you ARE. A dozy fucking CROW, mahn. Say it AGAIN.'

And the OR1 repeated it aloud two or three times before the veteran persecutor walked off, shaking his head sadly in acute apprehension for the future of the British army.

But being an OR1 meant that you were called to War Office Selection Board soon after the end of basic training at Palace Barracks. Beyond that, as many of us were already discussing and hoping, lay the possibilities of life in an officers' mess, with decent food, servants, a batman, better quarters, and the perquisites of rank. It was not so much that we wished to be officers as that we wished to escape from the bad food, the primitive living of mess hall and billets, and from guard duties.

As soon as we had passed off the square, we took our turn at night guard duties. The camp had two sets of night guards: one a fixed station at the main entrance to the camp, the other a roving 'picquet' guard inside the camp. Starting just before 'Lights Out' and ending at first parade next morning, they involved two-hour spells on guard, and four hours of rest between

each spell. On my first night of guard duties a fierce March rain-
storm buffeted the camp all night long. In very bad weather, the
sentry at the gate was allowed to stand just inside the sentry
box, provided he emerged at least once every five minutes to
patrol his 'beat'—the short stretch across the main gate which had
been my first glimpse of Palace Barracks early in January. Marching a few paces alone in blinding rain at three a.m. on a black
winter night, safely behind a tall, iron gate with spikes across the
top, seemed a lunatic exercise. But the possibility of enemies
within the gate was not entirely fictional. For many years the
Irish Republican Army had attempted occasional night raids on
armouries. We kept our rifles stacked together in racks in the
billet, but it was a rigid rule that all rifles were secured in the
rack by a heavy chain passed through the trigger guard and
secured by a padlock to the steel rack. We heard stories of daring
coups by IRA guerrillas disguised as British soldiers.

The roving picquet guard, or picket guard as Part I Orders and
printed instructions called it, seemed a more sensible proposition.
After inspection by the orderly officer of the day, the picket guard
shed their boots and put on plimsolls for the rest of their duty.
At 'Lights Out' they were divided into pairs, given electric torch-
lights, armed with heavy knouts, and detailed to patrol various
sections of the camp area, with stealth.

I preferred this type of guard duty, with its freedom to roam,
possibly to discover the unexpected or to perform some positive
role by discovering intruders within the gates. On my first patrol,
towards midnight, I heard suspicious noises from the bushes near
the back of the NAAFI stores and quarters. Thefts from the
NAAFI were frequent, prompted either by extreme need, or
merely deviant behaviour, as some sort of protest against the system. I drew near the rustling in a thicket close to the NAAFI
quarters, then crept forward quietly, my skin goosepimpled. I
jumped around a bush, raising my knout and snapping on my
flashlight. The beam revealed an NCO prone, with his trousers
down, straddling one of the NAAFI girls beneath him. Both
were extremely surprised, and then very angry. I switched off my
torch when I saw the two stripes on the soldier's tunic. The

corporal uttered the familiar expletive in the form of a command.

After initial weapon training and a good deal of repetitious 'firing' with dummy rounds, we were taken to the firing ranges at Ballykinler, about thirty miles south of Belfast on the eastern shore near Dundrum Bay. Memories of the days at Ballykinler ranges are chiefly of long waits at the firing detail, blowing the fingers to keep some sort of circulation going in the bitter February air by the sea. And of wiping the eyes to catch sight of the tiny distant target, sometimes four or five hundred yards distant. Or doubling across treacherous sliding shingle with a Bren gun and four magazines bumping about inside the pouches strapped to the chest.

The firing ranges also gave unique opportunities for the warrant officers to inflict special humiliation on ORls intimidated by the bangs, or by the kick of the rifle and the general alarms and confusions of 'rapid fire'. Each detail of about twenty men had to run down the range, firing from distances of five hundred, four hundred, three hundred yards down to fifty yards. Every man had to keep abreast, for fear of letting loose a stray round in the rear of his fellows. It was quite a spectacle to see elderly warrant officers chasing the weak and the timid, clouting their heels with heavy blackthorn sticks as they stumbled with Bren guns on the shingle, their spare magazines scattering with a fine, tinny sound on the pebbles.

'On your FEET, Jones 211. On your FEET you idle bastard, get GOING will you, you'll have the next DETAIL firing up your ARSE, don't you know. . .' And Jones 211 would come panting up to the next firing point, half sobbing with fatigue, scattering his magazines about him as the Bren gun bipod collapsed and the barrel dug into the sea turf. At the one hundred yard firing point Jones lay prone, his body heaving with exhaustion, but there was no resting. The next detail were already lining up at the six hundred yard line—mere specks on the distant turf. Two sergeants helped him to his feet, and the warrant officers concealed any concern behind the usual banter. 'Come on, then, Private Rigger Mortis. I bet you'll perk up when it's NAAFI break; won't you now?' Jones smiled wanly and apologised with public-school courtesy.

At Palace Barracks our final week of training ended with one or two lectures by the company commander on 'The Art of Leadership', at a fairly unsophisticated level. There were also two or three visits to the camp cinema, where we were shown gruesome technicolour films on the effects of venereal disease on the male organ. The films were prepared for the American army, the warning commentary spoken in a stern American basso profundo. The films reminded us of the hazards of sexual pursuit in foreign countries, although, at the end, our tall platoon commander in red Parachute Regiment beret warned us from the stage that we could just as easily 'get a dose' in Belfast or London as in Aden or Hong Kong. If we did, he warned, then it was not only self-interest to report to the camp doctor immediately, but also army regulations to do so. We marched back to company lines with a new concern for genital health.

Sex in its various manifestations was woven deeply into the lore of barrack life. Apart from the expletive in everyone's mouth, earthy jokes were bandied about over beer and in the billets. The more extrovert members of the platoon would cheerfully call across several bedspaces to their friends, passing on 'new' jokes, with the required gestures with boot brush or Brasso tin.

'Hey, mucker, d'ye hear the one about the NAAFI girl who wanted a corkscrew?...'

The more sensitive recruits pretended not to notice the graphic dénouement when it was reached. Others grinned appreciatively or guffawed at the right moments. One could marvel at the sheer picturesqueness and sense of imagination shown in these stories of gladiatorial sex. As the weeks went by and familiarity inside the billet leavened all conversations, the attractions of particular NAAFI girls in the camp, or of girls serving in the NAAFI club in Belfast, were described freely, with advice or comments on their availability and performance.

Once a week a salvo of sexual comment would break out in the billet when our laundry was delivered. Our army shirts, underclothes, socks and pyjamas were sent to a laundry in Dona-ghadee, at the army's expense. There the laundry girls had formed the habit of slipping brief notes into the laundry bags of some soldiers when they were despatched for return. On two or three occasions I received a *billet doux*, smuggled inside a sock the first time and then, more intimately, pinned inside my pyja-mas at a crucial spot. The notes often gave no more than the first name of the girls who importuned, as though courage had deserted them. But one or two dared to add their addresses, with the brief plea 'Write me a letter saying what you're like', or suggesting meetings on the seafront at Bangor.

Bangor, the little coastal resort on the shores of Belfast Lough, was the favoured place for picking up girls. The promenades and seafront cafés were suited to chance, or deliberate encounters. On Saturday and Sunday afternoons, in their twos and threes, the girls came in from the surrounding towns and villages, some no doubt from the laundry at Donaghadee, to meet the soldiers.

Our training ended with a flurry of tactical exercises, assault courses, and a final round of live firing on the Ballykinler ranges, using all the infantry weapons: rifle, Sten, Bren, mortar, PIAT and the grenade. We were now so fit, so ravenous at meal times, that any initial horror at the mess hall and the grease-filled troughs had vanished. We bolted our food, and in the evening filled up at the NAAFI canteen or at Sandes Home. We slept soundly on our rough beds under coarse blankets, without sheets, about twenty to a billet. In all, we accepted, even partly enjoyed, conditions certain to cause a major riot in any modern prison. Contentment, like aesthetics, is a set of attitudes within a context.

In the twelfth week we had our Passing Out parade, and dis-covered that we were now trained soldiers. Baker Company lines emptied once more as we left for a week's leave.

E

CHAPTER 6

Helen's Bay

We each received our next posting before going on leave. Many were to join the regiment of their choice, if it was based in the UK, whilst others went to their regimental depot to await the call to the War Office Selection Board, as our OR1 status allowed. Some were posted to the Officer Cadet Company of the Twenty-Eighth Training Battalion, at Helen's Bay, farther along the shores of Belfast Lough. We had heard a good deal about the pleasant, leisurely life there from reports carried to us by officer cadets returning to Palace Barracks to have a drink with school friends doing basic training. I was glad to hear I was among the Helen's Bay contingent from our intake at Baker Company, especially as my closest friends were also posted there.

Spring arrived early. The little camp nestled invitingly under the spruce and fir trees dotted about the promontory at Helen's Bay. The billets were superior to those at Palace Barracks—built much later, and more solidly, with a good deal more air and light. They were even sited with some eye for landscaping, dotted individually under the fir trees with little regard for geometry. There was a small square in the middle of the camp, but it was not hallowed ground, and no drill monster lurked at the edge or beneath the trees. The view across Belfast Lough was splendid, and our daily schedule was sybaritic after the Palace Barracks regimen.

As 'potential officers' we had several minor privileges. Weekend passes were encouraged, rather than rationed. There was no official 'Lights Out'. We wore the badges and shoulder titles of our selected regiments, with round white discs behind the cap badge, and white tabs on the epaulettes. We were allowed to wear plain clothes from lunchtime onwards, as parades finished at noon. Instruction was given at a leisurely pace—what little there was. Since we were now fully trained in weapon-handling, map-reading, and other basic skills, the instructors at Helen's Bay could do little more than give us more of the same.

The two or three officers in charge of us were more interested in sports or womanising than in maintaining discipline. The young adjutant, a staggeringly handsome man straight from some film set, often appeared with his girl friends in his open sports car, sweeping around the little square, returning our salutes with a genial wave, combined with half a smile and half a leer as he bore his cargo off to his quarters. The CO rode to hounds and was rarely seen.

As though to compensate for the absentee officers, a proficient, immensely tall Irish Guards sergeant was in charge of drill. We learned that he was something of a legend even in the Guards. There was first of all his sheer size. Seven foot tall, his battledress really did seem to be stitched on, with every ripple of the biceps and forearms showing as he stalked among us during our rare drill periods. His Guards cap, too, seemed to be part of his head, a mere continuation of his neck at the back, and apparently cemented to his brow at the front. A vestigial, vertical peak shaded his eyes. He was the prototype Giles cartoon guardsman. His main job was to get our foot drill up to the standard of the Officer Cadet Schools at Mons Barracks, Aldershot, and Eaton Hall, near Chester. Reputedly, the standard there was much higher than in any basic training unit. Sometimes we drilled with our officer cadets' canes, and this brought new flights of obscenity from the drill sergeant. He was a witty giant, with an Ulster brogue, attuned to the sensibilities of the Helen's Bay officer cadets.

The general air of levity and unseriousness in the camp carried over into our drill, and sometimes a cadet would deliberately prod the man in front of him with his cane as we drilled on the tiny square. The drill sergeant had seen it all before, scores of times. He would halt us, then approach slowly, ominously, to stand just behind the culprit. The Irish brogue was now, if anything, accentuated. He almost certainly knew that he was mimicked, with affection, in the billets under the trees.

'Did I see you poke this officer cadet up the rear orifice, now?'

It was best to admit the offence, as the sergeant did not like to be deprived of his prepared jests. He was even ready to overlook the smiles about the ranks, provided no one laughed out loud as the comedy developed.

'Yes, sar'nt, it was an accident, sar'nt...'

'An accident, now, phwas it? An' I saw you do it three times, didn't I? Well now, you know that sodomy is a serious offence for Other Ranks? Won't you not wait till you're arficers for that? You want to damage this officer cadet's prospects do you? How's he goin' to get on in the Hussars [or the Lancers, or the House-hold Cavalry, according to his shoulder titles] if people like you damage his rear orifice? You want to ruin his promotion pros-pects do you?'

It was good stuff for public-school men, and shoulders shook with suppressed laughter as the sergeant stalked away to resume his barked commands from the side of the square. He quickly be-came, in the customary way, a father figure, and he clearly en-joyed having an admiring family paying him his due respect on the square.

Although his main duty was to bring our foot drill up to stand-ard, a system of junior under officers, appointed temporarily among our own ranks, took care of some minor parades. Each morning after breakfast there was a perfunctory 'breakfast roll call' parade, with an inspection in order to satisfy the universal custom of the British army. The under officers formed us up, and once this was done, the dashing young adjutant might appear at the door of the small officers' mess across the square, and come forward to inspect us. On some mornings, when the under officer had us ready standing to attention, facing the officers' mess, min-utes would elapse and it would become clear that the adjutant had not yet emerged from the pleasures of the night, if indeed he was in camp at all. The under officer would then inspect us him-self, or make some show of doing so.

Even when the adjutant took the inspection, it tended to be a frivolous affair, his mind clearly elsewhere. One of us took a bet in the billets that he would be able to clean his teeth on the parade ground during the adjutant's morning inspection. I agreed to be one of the scrutineers and stationed myself next to the man who took the bet, in the rear rank. As the adjutant passed along the front rank, his handsome blue eyes blinking wearily, our betting man in the rear took out a toothbrush, squeezed some toothpaste, and brushed vigorously for a few seconds before putting away the brush and wiping his mouth. He won his bet.

In mid-April I was called to the War Office Selection Board at Chester and travelled there with several friends—some from our intake at Palace Barracks, some from other concurrent intakes. One of the advantages of Helen's Bay Company was that a good deal of first hand information was brought back from WOSB, and handed around freely. We knew roughly what to expect, even though there was the knowledge that, out of every six cadets summoned, only four or five at the very most passed, sometimes less. We were therefore competing against each other, and this introduced a nervous element as we travelled, each privately appraising the others.

Our nervousness was also compounded by reports that at some moment during our three days at WOSB each of us would be deliberately confronted by a totally unexpected situation, designed to catch us off our guard, and thus test us more realistically in the emergency conditions officers might have to deal with at times during their careers. Some of the reports brought back to Helen's Bay sounded very improbable; perhaps sadistic inventions by candidates, but we could not be certain. One story had it that at the end of an interview a candidate would be told to go through a door and take a seat in the room at the end of the passage. When the candidate opened the door at the end of the passage he found himself confronted by a major-general, wearing hat, Sam Browne belt, and field service tunic, sitting on a toilet seat with his trousers down. The test was designed to separate those who retreated in hasty confusion from those who promptly saluted the enthroned general, perhaps apologised, and then withdrew in good order. Some insisted that the story was a complete invention, the apocrypha of inventive hoaxers. Others said it was true, and a proper test of stoic qualities. When we arrived at Chester none of us was certain.

We were quartered in comfortable billets at a small camp not far from Chester, with the unusual luxury of bedside lamps and bathrooms. We were about thirty in all, assembled from training units in various parts of Britain, all of us designated ORIs by selection procedures at our training units. We dined together around three or four round tables, and were joined at dinner by the many staff officers littering the place, each with the red tabs of the general staff on the lapels of his tunic. We did not need to be told that our table manners and conversation at dinner might count in the final result.

After dinner the commandant, a major-general, told us what they were looking for in the series of tests, interviews and 'situations' of the next few days. This boiled down to 'qualities of leadership'. The general did not define these qualities. One saw the point of this omission. After his talk we were invited to have drinks at the bar in our allotted 'mess', and again the officers joined us, obviously ready to drink as much as any of the candidates. But we were forewarned of this initial hazard, and everyone developed a prudent abstinence as the evening wore on.

Next morning we were divided into groups, five to a group, and taken to a number of 'situations', mostly involving the transporting of a large object across some moat or trench, or over a high barrier by means of manpower and various items of equipment. Each of us in turn was put in 'command' of the other five members of his group, and firstly taken aside to be told the task, then allowed to inspect the available equipment and the hazard to be negotiated. After this, a few minutes to think, and then to issue the orders to the group.

During the next hour or so we performed various extraordinary feats, swinging heavy girders across moats by means of ropes or tripods constructed from railway sleepers and iron bars; or huge barrels negotiated along narrow planks over ditches, or over brick walls constructed with a fiendish absence of footholds. Some of the appointed 'leaders' of a group failed disastrously during the allotted time. The officers watched us, with clip boards

poised, taking notes constantly, clearly familiar with every possible approach to the problem. They quickly rescued morale with
a resounding 'Hard luck: nearly did it...' as the time ran out
and the chagrined 'leader' looked upon the wreckage which
marked the ineptitude of his 'plan', or worse, a palpable failure
of leadership.

There were other sorts of test to make up lost ground. Interviews were a recurrent phenomenon—some conducted by battle-
scarred majors, others by clean and clinical-looking men who were
probably psychologists. Their questions revolved mostly about
our reasons for wishing to become officers; our schooling, hobbies, aims, ambitions, and so on. We were also told that each of
us would give a short 'lecturette' to about twelve of the others,
with our own choice of subject. It would be followed by 'question time'. This involved some nervous or slightly embarrassing
exegesis as each candidate took his turn.

Candidates expounded their hobbies, such as sailing, or stamp-
collecting, or the delights of photography, and, in one case, the
pleasures of hunting with the Quorn. A tall Etonian from Helen's
Bay Company gave a fascinating talk on the sport of falconry
and clearly scored heavily with the officers present. We could
select any topic or subject we preferred and it did not have to be
a hobby. It could even be an interesting book. I gave a short
talk on the character of Bismarck, who had interested me at
school as statesman and militarist. It seemed to go down better
than I had expected, even though I merely repeated the gist of
a school essay on the subject.

After a final interview for each of us, with all the officers spread
around a long table, we waited to be told the results back at our
'mess'. About twenty-four had passed, the rest had failed. I was
relieved to learn that I had passed, and joined in the commiser-
ations for those who had failed. They knew that they could try
again in perhaps a few months' time, but there was no mistaking
the dejection, and in one or two cases the misery, of those who
would now return to the dreary life of soldiers' billets and NAAFI
canteens.

Back at Helen's Bay we awaited our call to Eaton Hall Officer
Cadet School near Chester, not far from where we had attended
WOSB. It seemed an odd waste of public money to return us to
Northern Ireland to await our posting back to Chester, within a
matter of days probably. But the pleasant life by the shores of
Belfast Lough in late spring despatched these thoughts.

During these dogs days one of the few solutions to our lack
of employment consisted of 'initiative tests' in the surrounding
countryside, with a few days under canvas at a place with the
romantic name of Clandeboye. The initiative tests were meant
to be competitive, but were not regarded very seriously by anyone. We were taken on some undeclared 'mystery' route in Ulster
in a covered three-ton lorry, no chink of light penetrating, then
dropped individually at various points with compass, map, rifle,
a packed lunch, and some written instructions. These included
brief directives to discover items of information, such as the time
of the last post out of Donaghadee; the names of the two pubs
at another small township; any six names from headstones in a
tiny graveyard some miles to the east of the same location; and
whether a certain stretch of land beyond another village was
suited to landing gliders for any possible airborne attack. When
all our items of information were gathered we were to rendezvous
at Clandeboye, not far from Helen's Bay, where we would find
the officers and our redoubtable Irish Guards sergeant already
under canvas.

The rules were that we were allowed to use any means of transport except public transport. With perhaps thirty miles to cover,
loaded with a rifle, dummy ammunition, steel helmet, small pack,
water bottle, gas cape and mess tins, the rule against using public
transport was thus a fairly constricting one in these years before
the automobile explosion and an end to petrol rationing.

I was put down on some unfamiliar country road. Very few
cars passed me as I set out with compass and map. Then a 'baby'
Austin car chugged around a corner, going in my direction.

I thumbed a lift. Happily, the driver stopped. It turned out to
be a small van, cluttered to the roof with bric-à-brac. The elderly
bent man inside told me he was going north to Donaghadee,

though he would need to stop once or twice on the way. I said that any sort of lift would be welcome. He bade me get in but this proved something of a problem. Inside the van a great variety of pots, pans, pails and hardware was piled, spilling over to the front passenger seat, which was well forward and could not be moved back. After a good deal of manoeuvring I ended up lying along the mound in the back, face upward, with my boots reaching forward across the front seat to touch the dashboard, my rifle clasped down my front from waist to ankles. We set off, my supine form like a crusader's tomb in some cathedral apse, my rifle clasped to my middle like a sword. I lacked only a faithful hound to cushion my feet and a wife in effigy at my side. Even in this recumbent and unready position, however, with my steel helmet biting into my neck, I was grateful to the eccentric old man, who seemed to forget about me as soon as we set off.

Unfortunately, it was soon clear that he was to stop at every house, cottage and hamlet along the road. I lay there, increasingly uncomfortable as he fumbled in the back of the van, and realised that I had accepted a lift from a motorised tinker, who liked to pass the time of day with each of his customers. Now and then he pulled a pail or a pot from beneath me, apologising when another iron pot fell sideways from the heap, dealing me another blow in the ribs or on the shoulders. My predicament was an embarrassing one. The old man knew my destination, and would be offended if I decided to leave him now. If I decided to walk, his present rate of progress would cause us to pass each other repeatedly on the road, with embarrassment to both sides. The situation called for initiative, if not diplomatic skill. I invented some story at the next village and disappeared into a church until he was out of sight. My next lift on a milk lorry was more speedy.

By mid-afternoon I had carried out my list of commissions and reached camp at Clandeboye. It was May, and the sun was warm as we sat about the grass holding post-mortems on our excursions, discussing map-reading techniques, the correct type of valley for airborne invasion by glider and by parachute. The easy tempo of life at Helen's Bay infected everyone, from the adjutant to the cooks. We all ate together on the meadow where we camped. We

were on the Clandeboye estate, a private demesne, though with
some special dispensation for Helen's Bay Company. Somehow,
the name Clandeboye was vaguely familiar to me, but I could
not be sure. As we talked in the sunshine, the young adjutant
mentioned the family of Dufferin and Ava. This struck another
chord of memory. Then he gestured to where the wood rose in a
gentle knoll. 'In those woods there, you'll find Helen's Tower if
it interests you.'

When our work was done for the day I slipped off and went
through the wood where the path rose upwards through beech
and chestnut. Then it levelled off, and beyond a curve I saw the
slim tower rising above a silent wood from the smooth green turf.
The setting was perfect, like a Doré drawing, and I now remembered the book I had read years earlier, at a time when I devoured
any biography I could lay my hands on. It was Harold Nicolson's
memoir on the Dufferin family, *Helen's Tower*. I remembered
some of the tale of love and despair, romance and tragedy which
lay behind this graceful folly in the woods, and particularly a
couplet Nicolson had used in his book, taken from Scott's
Rokeby:

> And now the stranger's sons enjoy
> The lovely woods of Clandeboye.

This was the couplet which had rung in my mind for most of
that day. The caretaker was cutting the lawn and when he asked,
'Have you come to go up the tower?', I took up the offer immediately.

He gave me the keys and I went up the stone steps. At the top
I came into a hexagonal room furnished with sombre antiques,
a deep red carpet on the floor and red brocade elsewhere. In the
centre of the room a hexagonal table was inlaid with compass
marks and motifs, indicating the various sights at the surrounding
windows: Strangford Lough, the Antrim Hills, the Mourne
Mountains—all visible in the fine clarity of the evening sun.
Nearer at hand, the woods of the Clandeboye estate clustered
beneath the tower.

Most fascinating of all were the poems inscribed on the walls
inside the room. One was by Browning, another by Tennyson,

memorialising the beautiful and tragic Helen, Lady Dufferin, for whom the tower was built, but who died of cancer only a few years after its completion. Disraeli was among her admirers. Other statesmen had courted her; some were favoured. There was a poem on one wall by Kipling. Up to this moment, Kipling had made little appeal to me. The heavy thumps of the metre in Barrack Room Ballads had always seemed unsubtle, close to doggerel. But now, standing among the dusty brocade, with the dark, wine-coloured damask at the window filtering the evening sun to pale russet across the silent room, Kipling's ballad had a sudden Gothic resonance. The verses were from 'The Song of the Women', written in memory of Lady Dufferin, partly as a tribute to her beauty but also to her courage as her life ebbed.

There were several verses from Kipling's poem, but two stayed in the mind as they had the elliptical, slightly unfocused quality of good poetry, with feeling beyond the limitations of ordinary language:

> Go forth across the fields we may not roam in,
> Go forth beyond the trees that rim the city
> To whatsoe'er fair place she hath her home in,
> Who dowered us with wealth of love and pity.
> Out of our shadow pass and seek her singing
> 'I have no gifts but Love alone for bringing.'

> By Life that ebbed with none to staunch the failing,
> By Love's sad harvest garnered ere the spring,
> When love in ignorance wept unavailing
> O'er young buds dead before their blossoming;
> By all the grey owl watched, the pale moon viewed,
> In past grim years declare our gratitude.

When we returned to Helen's Bay camp the following day we heard that the Officer Cadet Company was to be closed down. No one seemed to have had any advance news of this, but some new army order decreed that in future all recruits, whether future leaders or not, would carry out basic training at their brigade depots, according to the infantry brigade or army corps they wished to join. The new order meant that the Twenty-Eighth Training Battalion at Palace Barracks, together with the Officer

Cadet Holding Company at Helen's Bay, would cease its function
of training selected future leaders.

The saddest person was our sergeant in the Irish Guards, who
freely admitted to us, off-parade, that he had no wish to go back
to the Guards Training Depot. Each of us received posting orders
to join our regimental depots, where we could await our postings
to Eaton Hall Officer Cadet School. There were one or two nos-
talgic gatherings in our small NAAFI-cum-mess; mournful occasions partly cheered by the thought that we would meet again at
Eaton Hall. We made our farewells to the drill sergeant, and had
the awe-inspring experience of his handshake, a huge, bone-
bruising vice. Then the small community beneath the silver firs
and spruce trees on the shores of Belfast Lough dispersed. My
own posting was to Fort George, Inverness, the regimental depot
of the Black Watch. Despite some clinging nostalgia for Helen's
Bay, I was glad to be going for a first taste of regimental life.

CHAPTER 7

The Black Watch

I took the steamer to Glasgow, then the train to Inverness through the Cairngorms. The peaks were still speckled with snow. The journey was a curtain raiser for Fort George, which lies at the tip of a lonely promontory, staring across the Moray Firth—a superb defensive position in itself, the citadel doubly secured on a bastion of solid rock, surrounded by moats and thick granite walls.

I reported, but no one seemed to be expecting me. My posting orders showed that I had passed WOSB and was awaiting a call to Eaton Hall. The clerks in the cavern-like rooms in the fort were at a loss what to do with me. I was supernumerary in every sense of the word and my status of officer cadet conferred an uncertain rank. Eventually, army regulations rescued the clerks, as they usually do over such uncertainties in military life. After much casting about in the red book they established that even an officer cadet at OCTU is no more than a private soldier, from the point of view of rank and privileges. Accordingly (and with evident pleasure from the clerks) I was told to collect bedding and move into a vacant wooden billet just across the moat and outside the fort. It was much less impressive outside the walls of the fort, but there was the compensation of sea water lapping at the shingle only ten yards away, and a long and splendid curve of beach which gave fine views across the water to the mountains of Ross and Cromarty.

The other two or three men in the billet were old timers, in a literal sense. A soldier next to me had just been released from a long sentence in military prison for stealing quartermaster's stores. His previous rank had been quartermaster sergeant, but he had fallen for the temptations of the black market in Germany before being 'put inside' for about a year. He seemed to have no kit whatever, and freely borrowed everything I possessed, from shaving brush to razor blades and boot polish. I hid my tooth-brush, as he did not have the habit of asking permission first. But he seemed a friendly, harmless person, unless he was simply cowed by his prison sentence. Using my kit may have been one way for him to re-establish contact with the outside world. He borrowed the price of twenty cigarettes and did not return the money, but I could not think of it as a bad debt.

The number two cook-house, which served meals for the variety of unposted men outside the walls of the fort, was unbelievably repellent. Today, its conditions would probably cause a serious mutiny. Flies infested the tables, and sparrows hopped almost to the edge of our cracked plates as we tried to stomach the appal-ling fare served up. The men shovelled in the sodden cabbage and gristle which passed for lunch, each mouthful accompanied by grunted obscenities. The cooks had clearly long since lost any interest in food preparation, or regarded their present posting as some sort of punishment station, and took it out on those they were meant to feed. At each mealtime we were meant to 'fill up' with bread and margarine. The provision for this was made on trestle tables, where bread was laid out, two slices each, an inch thick, stale and curled at the edges, with a smear of rancid mar-garine decorating the top slice. You took your two slices—if you could stomach them—from the end table. If any slices remained at the end of the meal they stayed there for the next meal or meals. Some of the slices developed astonishing shapes as the days passed, curving upwards until the edges almost met in an arc.

Having installed us, the authorities in the fort clearly wished to have nothing more to do with us until fresh posting orders arrived. We ceased to exist. I made myself scarce each day. I now knew enough about the army's ways to realise that hanging about with no visible employment simply invited a spell on fatigues. So each day I went off to Inverness, walking about the town, swim-ming in the covered swimming pool there, or reading in the austere public library. Usually an army lorry would pick me up along the road, although one day I thumbed a lift from an approaching station wagon. This turned out to be the commandant, a friendly man who was himself off for a day's fishing to some distant loch or stream. He said that a battalion of the Black Watch were expected back from India next day, and told me I should make certain to see the parade to welcome their return to the regimental depot, pipes, drums and all.

I was glad I did. It was a brave sight. Inside the fort, on the smooth lawn before the officers' mess, with high castellated walls framing the scene, the pipes and drums marched and countermarched as the sun went down over the Firth. The men of the returning battalion stood out from the onlookers with their deep suntan and sun-bleached bonnets. Many of them looked like real veterans, their medal ribbons sprinkled with unfamiliar colours and orders. India and the death of Empire skirled about the fort. I felt at last that I was beginning to get a little closer to what the army was ultimately about. The pageantry on the enclosed turf lawn, with the regimental flag fluttering high from a granite tower above the sea; warrant officers and officers in their dark green and black kilts, the officers with long sticks like shepherd's crooks, the brilliant scarlet hackle of the Black Watch in their bonnets, and skean dhu daggers in their stockings as they leaned on the crooks like shepherds or lairds—all this marked an identity, even a sense of family.

If they had nothing else, everyone there had common professional standards, a visible self-respect, a tangible pride. It may be that somewhere at the back of my mind there was also a sense that these men carried, however unconsciously, a sense of history. Whether or not they realised this was beside the point. I was eighteen, my imagination was easily touched, and it seemed to me that, political implications aside, this private pageant on a lonely Scottish foreland, with the battle honours on the Black Watch colours fluttering above the sea-strand breeze, the setting of dark granite and green lawn, all made their links with the past, with similar scenes on this same spot as men returned from hot lands where the sun scorched, where life held few compensations. Whatever the politics of colonialism, these were men with professional standards of high calibre.

My posting to Eaton Hall arrived within a few weeks. The orderly corporal told me to collect a rail warrant from the fort, and by early afternoon I was at Inverness station, my kit bag packed, with steel helmet topping if off, the drawstrings pulled tight about the rim. The journey through the night was long and tedious. I made the ritual change at Crewe and arrived at Chester the following midday. About a dozen other soldiers were on the train, quite different from my Fort George billet companions. They had smooth complexions, and different accents. Some of them sounded hearty and confident as they swung their kit bags and summoned a porter. A three-ton lorry awaited us. On the way to Eaton Hall we eyed each other carefully, no one speaking. There was a mixture of regimental shoulder titles and cap badges: Middlesex, Royal West Kent, Royal Sussex, Lancashire Fusiliers, Inniskilling Fusiliers, one or two light infantry regiments.

I had heard reports at Helen's Bay of the architectural oddity of Eaton Hall. A residence of the Duke of Westminster, its style a mixture of neo-Gothic and Scottish baronial, the hall had been acquired by the army to house a wartime Officer Cadet Training Unit. The duke returned now and then to occupy a small wing but apparently he preferred sunnier spots in various parts of the world.

We approached the hall along a perfectly straight road, the customary avenue of trees cut back to frame the hall behind what were called 'the Golden Gates'—vast ironwork pieces, heavily ornate, painted black, but tipped with gold paint at suitably embellished points among the spikes and cherubs. To the left of the courtyard (marred by temporary wooden huts), a private chapel stood slightly apart, though linked with the main structure, its small clock tower a miniature of Big Ben. Evelyn Waugh's description of Brideshead came to mind.

Drill sergeants from the Brigade of Guards festooned the place, all of them looking terrifyingly competent, some with waxed moustaches, all with bright eyes. They looked us over silently as we clambered down from the lorry, their pace sticks horizontal under the armpit, but clearly relaxed; whether with relish for what was in store for us was not clear. We were taken into the hall by officer cadets with the scroll-like motifs of junior under officers at the cuffs of their sleeves. These JUOs, we learned, would be commissioned as officers within a few weeks. As at Palace Barracks, several intakes were at various stages in their training and the best among the senior group were selected as under officers to provide a basis of self-discipline among the officer cadets. The best cadet in each company was senior under officer, with three platoon JUOs to assist him. Some of them proved to be much fiercer, even more sadistic, than the drill sergeants.

A JUO in the Scots Guards checked our names and then led us up the broad marble stairs inside the hall, past a baroque organ and one or two pieces of statuary. We were quartered in the state rooms, each containing four army beds, four wardrobes, four bedside tables, four lockers. It was very different from Fort George. The huge open windows gave out to terraced gardens, ornamental ponds and yew hedges. The sound of tennis reached up from the lawns. Two adjoining bathrooms were enormous affairs, with outsize baths marooned like galleons on platforms. The wash basins, set in marble, were broad, deep, capacious, with ornate taps.

The three officer cadets with whom I shared a room belonged to English county regiments. As the days passed I felt I was unlikely to make any new friendships among them. Each was from a minor public school. Perhaps they over-compensated with so many Etonians and Wellingtonians about the place. There were plenty of clues to a man's school. We were allowed to wear plain clothes outside parade hours, and the general atmosphere encouraged the wearing of school ties, especially when the officers invited us to cocktails before dinner.

The training was intensive, and the drill much harder than any of us had so far encountered. Those who had trained at the guards depot found it much the same, they said, but at Eaton Hall the permanent staff had the special threat of 'RTU' hanging over us at all times, quite apart from the system of punishment within the Officer Cadet School itself. The constant threat of being 'returned to unit' if one did not stand up to the pace, whether on the drill square or on tactical exercises, gave every-

F

thing we did a uniquely manic quality. Everyone tried to outshine everyone else in a race for survival. We learned—by the usual bush-telegraph of military communities—that some War Office regulation required the commandant to show a positive return of 'RTU'd' officer cadets at every intake, if only to show the establishment that he was maintaining standards. From the vulgar splendours of Eaton Hall, the horrors of life in the ranks could be recalled starkly, so that as the weeks passed, and the prize of a commission and thus permanent escape came nearer, the intensity became more marked. The drill sergeants were fully aware of this as they extracted a full measure of sweat and toil, and occasionally a little blood, on the square. Perhaps, too, they slightly resented the status and privileges the officer cadets would soon claim as commissioned officers. On the day we were commissioned, these same drill sergeants would have to salute us, and they could hardly be blamed for having this in mind when the weaker cadets floundered in their drill.

The rule was that drill sergeants and all members of the staff below the rank of officer would address us as 'sir'. This was presumably to accustom us to officerhood, to get us used to the habit of deference and to react to it properly. We in turn were required to address all members of the staff as 'staff', which was logical enough, but the system of mutual respect tended to produce curious verbal exchanges when we were being bullied on the square. The typical exchange went thus:

'Mister Smith, sir, you're a stupid, idle officer cadet. What are you, sir?'

'A stupid, idle officer cadet, staff.'

'That's right, sir, now wake up!'

Haircuts at Eaton Hall were a source of scornful wit for the drill sergeants. We were told to keep our hair short, but we also knew that the convict-like haircut of basic training was not really expected. It was equally clear that we were not allowed the length of hair associated with officer status; that is, curling out slightly at the ears, and touching the collar at the nape of the neck. We were betwixt and between. Vain officer cadets tended to try it on, tucking what they could beneath the beret and hoping that they

would get by for another few days. But once again the drill sergeants had seen it all before. Moreover, as Guards sergeants they
had to keep their own hair at regulation length—severely short.
At drill, therefore, a familiar gibe was the old cliché of standing
behind the officer cadet who was chancing his luck, the sergeant
beginning, *sotto voce*:

'Am I hurting you, sir?'

'No, staff,' in a tone of helpful innocence, anxious to oblige.

'Then I ought to be, sir, 'cos I'm standing on your HAIR. GET
IT CUT!' The last few words were bellowed close to the officer
cadet's ear, and it could affect his drill for some minutes, which
gave the drill sergeant more opportunities for scorn. But more
inventive sergeants did not fall back on the old cliché about standing on one's hair. They relied on fine metaphors about mistaking
the sex of long-haired cadets, or they recommended brands of
feminine shampoo, or powders and pommades. Or again, 'You
want to be commissioned in the Women's Royal Army Corps,
sir?'

'No, staff.'

'Well, sir, you're heading that way with hair like yours. Have
to draw you two pair of knickers, khaki, you go on like this. Look
at it! Hangin' on your shoulders, sir: officer cadets behind you
is twitching all over with desire. You got them deeply disturbed,
sir.'

Obscenities about promiscuity on parade were added, although
the canon differed slightly from that at Palace Barracks. There
was less about female pubic hair and pudenda, more about rear
orifices and sexual deviance. On the drill square, as in music halls,
the wit was geared to the audience. In this case, it was predomin-
antly public-school.

Regimental Sergeant Major Copp, of the Coldstream Guards,
presided over this brood of drill sergeants. Like all RSMs, his
status and station required him to intervene now and then in
order to show both drill sergeants and officer cadets who was in
charge and, at the same time, what abysmal drill a squad was
producing. RSM Copp lurked near a small wooden hut close to
the Golden Gates which protected the holy ground of his drill

square. He was getting on in years, and bulged from his Sam Browne belt, which flashed like a flexible mirror in the sun. But he seemed to lack that sense of humour which normally lightens a drill sergeant's minor sadism, and some of the officer cadets felt that there was a genuine streak of cruelty in him. It was difficult to say. He was taciturn and certainly looked humourless. His voice had a grating, piercing quality, and it seemed to be reserved entirely for admonition. He would stop a platoon of officer cadets drilling on the square, warn that if they did not 'pick up the feet' or smarten up immediately they would be marched up and down a deserted aerodrome about half a mile from the camp. 'Till the sweat shows on your backs and legs,' he would threaten.

Usually the threat was enough, but every platoon was eventually given a turn on the airstrip, and the RSM showed a remarkable crispness of pace as he marched us out at double quick time. The lonely airstrip was an intimidating sight—vanishing in the distance ahead. RSM Copp's voice carried well, especially along the airstrip, but once or twice he misjudged the acoustics, or the direction of the wind would change suddenly on the flat and empty landscape, and a bizarre scene would result. We would continue marching towards the horizon, waiting for the distant command 'About Turn.' A flurry in the ranks meant that some thought they heard the command, borne along on the wind. Others had not. In these circumstances, those who continued to march would barge into—and push along—the uncertain ones. A brief flurry and struggle followed. Curses were exchanged sibilantly in the ranks. But the marchers usually outnumbered the turners-about, since it was safer to obey an earlier order than to invent a new one, so we would continue marching. As we vanished in the blue haze of distance, the RSM would despatch a drill sergeant after us at the double. Perhaps this was all part of the process of preserving his authority—and the sergeant would relay the order, adding his own insults about washing out our ears next time we came on parade. Then the sprint back to barracks, at the double, rounding it off with five—or was it ten—minutes of double mark time, the drill sergeants tapping our knees up with their pace sticks, until our knees came up chest high, the sweat trickling down our backs. Back in our rooms at the end of the drill we peeled off our drill order. Shirts, trousers, battledress blouses, socks; all were wet through.

CHAPTER 8

Eaton Hall

The training was designed to develop soldierly qualities of toughness, endurance, and physical courage, but equally to introduce the habit of command and qualities of initiative. The Tactical Exercise Without Troops, or TEWT, tested all these qualities. For this we were introduced to the military appreciation, a universal formula which at first seemed to me an inflexible, constricting device. But later, after we had applied it to various tactical problems and situations, it seemed well grounded in clear reasoning and was even intellectually appealing. Perhaps its strongest claim was that it resulted in a single, well defined decision which was prompt and unambiguous.

The definition in itself had clarity and rigour: 'An appreciation is a logical consideration of facts, in order to arrive at a reasonable plan, to attain a definite object.' This seemed economical and useful, given the immense variety of tactical or strategic situations it had to cover. The formula for an appreciation was quite simple. You first set down the bare essentials of time, place, and map reference for the locale. You then stated the object to be attained, in simple direct language. 'To destroy...' 'To take...' After that, 'factors affecting' were examined. These included only strictly relevant factors, such as relative strengths of enemy and own forces; armament; ground; distance; time factors; supplies; reserves.

The next stage was 'Courses Open' both for the enemy and for oneself. These might number two or three, but rarely more. Usually two were considered sufficient, and their relative advantages weighed. Then the course adopted, then the plan, which had to be clear, simple, definite, but allowing some degree of flexibility. The appreciation complete, the leader would issue a warning order, getting his troops ready, making sure that all preparatory action was concerted. This was followed by verbal orders, given before attack or withdrawal, with sufficient time allowed for full preparation and administrative detail. The intellectual structure was coherent, economical, and by any standards appropriate to its purpose. Constant practice trained the mind in the direction of decisiveness.

We took turns conducting a Tactical Exercise Without Troops; each first assessing a tactical situation, then issuing orders to hypothetical troops, supporting units and armaments. As at WOSB, staff officers were in constant attendance, their spring clipboards poised, taking notes vigorously as we gave out orders or answered questions from our temporary subordinates in the field. Once again the element of competition sharpened our wits.

Lectures also formed a large part of the instruction, dealing with military law, the Army Act, the rights, duties and responsibilities of officers, warrant officers and other ranks. Advanced map-reading, which included a certain amount of trigonometry and geometry, replaced the more rudimentary compass plotting of basic training. The standard of instruction was high, if only because the officers acting as instructors were clearly chosen because of their special gifts. They were confident, zealous, energetic, and likely to rise in the General Staff in future years.

When they were not giving forceful lectures, with a minimum of verbal ornament, or running around the Eaton Hall pastures with us on TEWTs, these vigorous majors and captains provided a novel form of instruction. This was the short 'playlet', written and performed by the officers themselves. Each 'playlet' pointed a lesson or provided an issue and this was discussed at the end of the performance. The officers enjoyed themselves in their parts, and latent talent often came out on stage.

The morals drawn were simple and direct. A favourite one was that of the young officer who liked to be chummy with his NCOs and men, using Christian names all round, and dropping in un-invited to the sergeants' bar. The dénouement came on active service when the young officer gave a command in a tricky situation and no one took any notice of him.

Most of the 'playlets' contained earthy military humour in the script, though the universal expletive was avoided. Sexual innuendo was the limit, forthright stuff which the Lord Chamberlain would have banned if the theatre had not been playing to a special private audience. At the end of the final curtain we would discuss the point of the play at some length, commenting freely on the characters and suggesting general lessons to be drawn from the action portrayed. It was a painless, amusing form of instruction which one was initially surprised to discover in a military environment. Within its own walls, the army was much more relaxed than the caricature in the outside world suggested.

Each of us had to give one 'lecturette' of about twenty minutes to fellow cadets. This was discussed and criticised by the audience, led by an officer who was known to be remorseless on bad lecturing technique. Like all such ordeals, one was a good deal better for having faced it and carried it through. The subjects chosen by individual officer cadets tended to reflect the extrovert interests of middle-class young men who had sailed, or collected stamps, or followed otter hounds, or otherwise led clean, open-air lives, well suited to prospective commissioned rank during National Service.

My own offering was on the Franco-Prussian war—and the military reasons for the Prussian victory. I relied heavily on some history essays from school. (I had already taken the precaution of having my essay books and notebooks sent from home to prepare, and a couple of visits to Chester library filled in further details.) I worked in a little Clausewitz, including his advice to the Prussian Crown Prince: 'Be audacious and cunning in your plans, firm and persevering in their execution, determined to find a glorious end, and fate will crown your youthful brow with a shining glory, which is the ornament of princes, and engrave your image in the hearts of your last descendants. . .' These injunctions seemed to me to have a lot to do with the events of the next hundred years, and I suggested direct lines of cause and effect from *Vom Kriege* to the victories at Metz and Sedan.

It was difficult to judge the effect on the audience, or, more importantly, on the officers. At the end, questions came up on the differences between the Prussian ethic and the way the British went about things. Why, then, had Germany lost both wars in the twentieth century? Privately, I thought that American industrial potential and manpower rather than superior strategy or tactics had much to do with those results, but it did not seem wise to stress this in front of a patriotic audience. So I took refuge in tales of British morale and the British tradition of losing every battle but the last one. Any lack of historicity in the interpretation went unchallenged. The presiding officer seemed to be happy about it, and, after his customary advice on the need to eradicate bad habits of delivery or audibility in the lecture, I passed this particular hurdle.

There was little time or opportunity for forming close friendships at Eaton Hall. The intensity of the atmosphere prevented easy acquaintance. To a degree we were all competing against each other. 'RTU' was a constant sword of Damocles hanging over our heads. Beyond that, there was a strict order of merit in the eventual list for commissioning for those who stayed the course to the end. Those who finished high up in the list could expect to be commissioned in the regiment of their choice, usually their county regiment. Some were trying for the Household Brigade or the Sixtieth Rifles. Places were limited, however, and if one's final tally of marks for drill, turn-out, qualities of leadership, lecturing ability and so on resulted in a low placing, then this would mean direction to one of the less popular regiments, or even to a corps, such as the Royal Army Service Corps, held in low esteem. In this sense, everyone was competing against his neighbour throughout every day.

One pleasant and casual relationship I formed was with someone who, like myself, listened to music on the common-room radio occasional evenings. We had both listened to Bach one evening, and, as the weather was warm, with sunshine streaming across the terraced lawns to the rhododendrons by the ornamental pond, we strolled through this unmilitary landscape, talking about heaven, hell and the hereafter. His name was John Blacking, and he was going up to Cambridge to read Theology, then

to enter the Church, and try for a slum parish in London's East
End. His view of heaven and hell was that in the afterlife all
human souls took the same journey to a common destination.
Hitler and Hannibal, St Francis and St Peter, Rasputin and Joan
of Arc—all took the same journey to the same Elysian fields. But
once there, everyone and everything was equal so that those who
had devoted earthly existence to others found complete satisfaction in continuing to eternity the mode of life which had brought
them contentment on earth. On the other hand, the corrupt, the
vain, the selfish, the cruel, found the same spot an everlasting
hell, from which there was no escape.

There were some difficulties about this eschatology, it seemed
to me, but it was more attractive than Calvin's dictum on the
souls of the elect. We argued it out as we strolled around the
ornamental lake with its water lilies and banks of rhododendron.
It was perhaps an odd conversation for two officer cadets to conduct on a warm summer evening in sight of the Gothic turrets
and miniscule spires of Eaton Hall, but it was the first real tête-à-
tête I had had since walking with Dick McWatters on Grey Hill
above the camp at Holywood Barracks. We said we would keep
in touch, although John Blacking was about to be commissioned
and I presumed later that the Church claimed him after Cambridge. In fact he had a change of mind, read anthropology at
Cambridge and turned up twenty years later as professor of social
anthropology at Belfast, and a leading scholar on the musicology
of primitive societies.

In the fifth or sixth week of training, our intake took its turn at
a three day battle camp at Bickerton, not far from Chester, where
we were under canvas for a simulation of battlefront conditions.
The area showed the scars of the many tactical schemes 'fought'
by our predecessors. Rough scrub land, with raddled trees, mostly
dead or dying owing to the ravages of thunderflash and gun
powder, and a good deal of bracken for cover. Topping it all,
the steep escarpment of Bickerton Hill, capped with gorse, and
ideally suited to a last stand by any beleaguered company of men.

We were split up into two forces—the traditional blue and

red, with arm bands to match. As the attacks were to take place
at night, the colours did not appear very helpful, but it seemed
best not to raise the point. Everything was taken very seriously.
The officers stressed the absolute importance of good battle discipline. No careless lights at night; trenches dug deep and properly
revetted, or shored up; firing positions exact and efficient for the
declared purpose, whether rifle, Bren or mortar; vehicle discipline
tight; camouflage against air attack. We were to appoint our own
cooks, signallers, patrols, guards, and so on.

By last light our blue force was dug in along a valley about a
mile from red force. On this first night we had to attack red force
(entrenched on top of the escarpment), take prisoners, and either
capture their position or neutralise it. One member of our contingent had already been designated commander of blue force
and after conferring with the directing staff officers from Eaton
Hall—hovering as usual with their clipboards and pencils—he
gave us his battle orders. His plan was not a complicated one. It
involved a sweep beyond the escarpment, diversionary fire from
a section dropped at the other flank, and then the main attack. I
was made leader of a battle patrol to go out shortly after dark
to cause what damage it could and give diversionary fire from the
flanks of the escarpment.

When the battle broke out Very lights, flares, and thunderflashes made a fine confusion, lighting the sky like day.

Next night it was our turn to be attacked, and as we lay in our
trenches at the alert, it came very much as we had expected. But
one unusual development brought a moment of genuine alarm.
The final charge of the attacking force over our positions was
carried out with fixed bayonets— part of the realistic simulation
of battle conditions. We were not expected to flourish our bayonets too closely. But among the officer cadets training at Eaton
Hall were one or two Africans, preparing themselves for commissions in the West African Frontier Force, the Kenyan Rifles, and
other African units. They tended to bring to all battle exercises
a special acrobatic gusto. On this particular night one African
in red force temporarily ran amok as the battle noises developed
with the usual thunderflashes, magnesium flares and the bangs of

dummy rifle fire, making a general hullabaloo in the night. Fixing his bayonet with a relish that alarmed his fellow bayonet chargers, he ran forward, leapt over the barbed wire well ahead of the field, and ran up to our trenches, yelling fiendishly under the bright flares. I saw the crazed look in his rolling eyes as he bounced forward, bayonet flashing. Fortunately I was not the immediate object of his charge. Instead, two thoroughly frightened public-school men in the next trench looked up with alarm at the leaping, hopping black man bearing down on them. 'I say, watch it there!' one of them shouted as the bayonet waved above him. 'Look here, cut it out,' the other one chimed in. There was a definite edge of panic in his voice. His partner was now cowering in the trench. Fortunately the African's fantasies deserted him at the last moment and he leapt across the trench, shrieking, towards the rear of our position. It was an awkward moment for British phlegm.

A final daylight battle took place that afternoon. The two forces met head-on in the field, minus bayonets this time, and there was a great deal of banging and cursing as men struggled to overpower each other and drag away prisoners. An oddly memorable incident came at the height of this battle. Everyone tried to show immense verve and seriousness as the officers watched eagerly from the side. We were very conscious that, locked in combat, individuals could earn a final spurt of marks for the merit list on the day of commissioning. At one moment the smoke of battle cleared and I found myself confronted by an amiable friend in the opposing red force. We flourished our rifles menacingly, then the billowing smoke cut us off from the watching officers. Observing this, with the smell of cordite still in our nostrils, the loud bangs of the thunderflashes and yells of rage all about us, my friend leaned languidly on his rifle for just a second and said quietly, cosily, 'Doesn't this all strike you as bloody ridiculous?' Then the smoke lifted, he grasped his rifle again, yelled like a maniac and charged off to the flank.

That night, with the battles over, we struck camp and marched back to Eaton Hall. Rain was falling steadily and we were exhausted through lack of sleep and the day and night 'battles'.

This final night march of about nine miles was meant to be an endurance test, carrying full kit and weapons. Some of the officer cadets suffered badly, and friends carried their rifles or parts of their equipment for them. We were covered with mud, and the sticky black grease of the camouflage on our faces hardened in the rain. There was no singing, and after five or six miles no rhythm to the march. Men who fell behind to tug wet socks away from blistered heels had to run forward again to catch up, their rifles and mess cans rattling from side to side as they waddled forward to find their place in the column. We stumbled into Eaton Hall about 3.0 am. The long, deep baths of the state rooms never felt so good.

First parade was at eight hundred hours. I discovered I had to travel immediately to Sussex for interviews at the Regular Commissions Board.

CHAPTER 9

Sandhurst Hurdles

Although most of the officer cadets at Eaton Hall were on National Service engagements, some of us were 'regulars'. This meant that we could either proceed to 'short service' commissions to complete a five-year engagement, or try for entry to Sandhurst, if we successfully passed the Regular Commissions Board.

I had agreed to let my name go forward to the Regular Commissions Board from rather mixed motives. If I continued at Eaton Hall Officer Cadet School, I would be commissioned in five weeks' time, with a short service commission for the remainder of my engagement of five years. Yet I was already finding many aspects of military life tedious. The few close looks I had been able to get of the officer's life confirmed that I would find it boring most of the time and even positively distasteful at times. When we took cocktails with the staff officers at Eaton Hall once a week before dinner, the conversation tended to revolve around set subjects: public schools, and their united front against the Labour Government and all its works; the attractions of county pursuits and county people; the cricket season; travel in hot countries; the White Man's Burden; and now and then outright racism wherever particular nationals could be subsumed under the compendious term 'wogs'.

Yet the obvious alternative to commissioned rank was even more repellent. My memories of the bleak billets of Palace Barracks, the odious mess hall at Fort George, were still sharp and pungent. Like every officer cadet at Eaton Hall, I had no wish whatever to return to life in the ranks.

The third possibility was to accept the call to Regular Commissions Board—if it should come—and perhaps then go on to the Royal Military Academy, Sandhurst. There, instructors taught courses leading up to BA standard, it was said, in such unsoldierly topics as the Modern Novel, French Literature, and Economics. I decided this was now the best among my choices. Several of the staff officers at Eaton Hall confirmed that post-war Sandhurst was completely different from the pre-war huntin' shootin' fishin' atmosphere; that it was now much more a place of education than a residence for idle young gentlemen seeking a military career in the British tradition of amateurism. Much of the teaching was by civilian dons, not staff officers.

The immediate choice thus lay between mouldering in an officers' mess—so far as I was concerned—and undergoing Sandhurst courses of instruction and learning. I preferred the second, so now found myself on a train to Sussex, with five or six others.

Our instructions were to go by train to Horsham, where a vehicle would collect us and bring us to Knepp Castle. There, the Regular Commissions Board would interview and test us over four days. The tests and the general atmosphere were similar to WOSB though more rigorous and searching. There were more staff officers lurking, with red tabs at the lapels, and, again, clinical-looking men at longer interviews on career intentions. The commandant was a major-general with rows of battle honours spread above his tunic pocket.

Knepp Castle was of uncertain architectural style, with mongrel castellation along the façade, giving it a faded fortress air. About twenty candidates were present in all, mostly from infantry regiments of the line, some from the Household Brigade, some Light Infantry, one or two gunners and others from the cavalry and the corps. We dined six or eight to a table, spread in a circle, with two or three staff officers mixed at strategic intervals. Once again we did not need to ask whether our conversation and general table manners would be closely observed during the next three days. Meals were thus transformed into punctilious and energetic displays of passing the salt at dinner, or the marmalade or toast at breakfast. Conversation was artificial and forced for the first two or three meals, then relaxed a little as the day's events provoked genuine, if limited discussion.

Nevertheless, conversation was one of the tests here. In the early evening, before sherry, we were gathered into two groups and invited to sit down in easy chairs in a loose circle, which presumably represented life in the officers' mess. We were then told to discuss whatever we chose. Any member of the group was free to initiate any topic: it was up to the remainder of the group to pursue it. Getting a conversation started was the main problem.

The staff officers sat discreetly to the rear, though obviously ready to make mental notes of every word said. The first few minutes tended to be like an uninspired Quaker meeting at the Society of Friends. The clock in the mess sounded unnaturally loud. On one evening when several minutes passed and no one spoke, a staff officer came forward and contributed a prepared, no doubt frequently used gesture of assistance. He threw a copy of *The Times* on to the carpet, which now stretched between us like a cultural desert. 'Talk about that, then,' he said.

We talked earnestly, praising the balanced reporting of political life at home and abroad, the sombre excellence of the leaders, and, in the case of one candidate, the full coverage of public schools' sports, especially squash and racquets. It was difficult to know what the observing staff officers thought of him, or indeed of our discussion generally. They gave nothing away. Whatever we did or said, the response was always the same: a courteous nod and a half smile, sometimes a terminal 'Jolly good', or a 'Well done'. Yet the staff officers were neither fools nor blimps. They struck me as shrewd men with unusual gifts for spotting strengths and weaknesses in the candidates.

By Friday evening we had completed all the outdoor tests of initiative, the indoor tests of intelligence, spoken and written, a variety of interviews, discussions, and the inevitable 'lecturette' from each candidate. On the Saturday morning, we were summoned individually before the assembled board, with the commandant seated in the middle. The officers and civilians appeared to have a great deal of information before them in multiple copies of a bulky folder. The commandant asked some questions, then said the board would recommend me for entry to the Royal Military Academy, Sandhurst. They smiled encouragingly. I was relieved that this particular uncertainty was now out of the way.

CHAPTER 10

Military Academy

The Sandhurst entrants were posted to a holding company at Mons Officer Cadet School in Aldershot. Here we did very little but read and go up to London at weekends. Then on a hot August day a small column of troop-carrying vehicles arrived from Sandhurst to take us to the Royal Military Academy. The big TCVs, dark green, highly polished and sparkling, carried the Sandhurst crest. We went through Farnborough and Frimley to Camberley, then slowed as we came through the white gates. The grounds of the Royal Military Academy looked very unmilitary at first glance: more Capability Brown than Prussian, although eighteenth-century cannon were placed here and there—black, elegant, highly polished. Then the long, low façade of Old College appeared across the lake, its white, two-storey façade and central Doric portico standing sharp and clean behind more cannon pointing across the parade ground.

In the next hour or so we followed the written instructions we had received for registration, collecting our Sandhurst passes, maps of the grounds and buildings, a cheque book for the academy's bank, where our pay would in future be credited, and finally our badges, leather belts, a lanyard, and two sets of white 'gorgettes' which the academy tailors sewed on to the lapels of our battledress blouses next day, along with our new shoulder titles.

All regimental badges and titles, together with regimental fads such as Fusiliers' plumes, the Highland Brigade bonnets, the Black Watch tartan and hackle, were discarded. We all wore the scarlet shoulder title 'RMA Sandhurst', with leather belts, berets, and the Sandhurst cap badge. The badge had the elegance of simplicity. At the base, the Sandhurst motto read simply 'Serve

G

to Lead'. At first I thought this was too plain, too cryptic. I preferred the motto of the Black Watch: *Nemo Me Impune Lacessit*.
But later I came to see the subtlety of the Sandhurst motto.

As at Eaton Hall, under officers buzzed about, crisp, alert, and
very professional. The staff officers seemed civilised and intelligent, mostly with the rank of major, here and there a lieutenant-
colonel. Many of them had decorations on their campaign ribbons. The warrant officers and sergeants of the permanent staff
were all Brigade of Guards, as we had expected. They looked in-
credibly efficient, disturbingly healthy.

We were talked to by various officers of fairly high rank, including the deputy commandant. We learned that there were three
colleges, with four companies to a college, making about a thousand officer cadets in all. Old College, where we were now
addressed, had been the original nucleus of the academy. Away
through the trees across the grounds we could see New College
and Victory College, built during the thirties to accommodate the
expanding numbers of officers needed by the army.

I was glad to find myself a member of Old College. We soon
heard that it was considered the most prestigious, in the way that
old foundations tend to be. One could admire its elegant white
façade and low profile, quite different from the modern mock-
Georgian rising through the trees beyond the cricket pitches.
Within Old College I was a member of Dettingen Company. The
other Old College Companies—Blenheim, Waterloo and Inker-
man—were also named after military battles where British soldiers had defeated the king's enemies. Our company commander
was a much decorated major in the cavalry. The platoon commander was a hereditary Scottish peer in a highland regiment.
He had a DSO and, as we discovered later, concealed a sharp mind
behind a languid air.

Each company was composed of three divisions: senior, intermediate and junior. The senior division would be 'passing out',
that is commissioned, in December, at the end of the present term,
and they enjoyed special privileges. Junior division had very few
privileges, if any. During the next few weeks we discovered that a
threefold weight of authority bore upon us in order to bring us

up to Sandhurst standards. At the top were the officers who gave military instruction. They lived within the colleges if they were bachelors, in rooms distinct from and very superior to the bare cells where we were quartered. These instructors marked our progress day in and day out, usually with two or three staring at us in lecture halls or model rooms as we discussed military tactics or the principles of war. They used the group photograph of the junior division (marked with our individual names) which was one of the first items of business on our day of arrival.

For drill and generally organising us for parades and time-tables, there was an abundance of Guards sergeants, with a Guards CSM to each company, and a College Regimental Sergeant Major. Above them, in the empyrean beyond mortal flesh, was Regimental Sergeant Major Lord, of the Grenadier Guards, master of all he surveyed whenever the academy drilled together. RSM Lord was a splendid looking man; tall, slim, a ramrod of a soldier, with a clear, cultured voice.

Like headmasters, RSMs tend to attract anecdotes. No doubt the tales serve as functional myths to unify the underprivileged group. The jokes which circulated about RSM Lord were affectionate. One joke said that when Lord's wife gave birth they were able to tell in advance that she would bear triplets, as the attending doctor distinctly heard a tiny voice within the womb, ordering, 'Get fell in, you two; father's going to inspect us as soon as we march out.'

The apocrypha were hardly necessary in the case of RSM Lord, since we had the legend before us when we drilled. He was too Olympian a figure to take the junior division on drill, until we had come somewhere near the level of perfection he demanded. But he briefly addressed us en masse, before handing us over to his underlings, the drill sergeants with their pace sticks and moustaches, all of them looking as though the demise of puttees had been a sad blow for general standards of turn-out. RSM Lord's voice came across the Old College parade ground crystal clear as he told us how to pay compliments to officers and to himself. He had a sense of humour.

'Gentlemen, my name is Lord, Regimental Sergeant Major

J. C. Lord, Grenadier Guards. So far as you are concerned, J.C. stands for Jesus Christ. When you're on parade, I address you individually as "Sir", and you address *me* as "Sir". The only difference is that you mean it, and I don't. When you're not on parade, you address me as "Mister Lord"—plain "Mister". But when you're ON parade, I'm not Mister Lord, remember, I'm Lord God Almighty. Now straighten up gentlemen.'

In fact there was nothing sadistic about RSM Lord. His introductory remarks skilfully conveyed the essence of his role and our relationship to him. Off parade, we would find him a courteous (as distinct from a deferential) member of the permanent staff at Sandhurst. But on parade the relationship was totally different, of necessity. Trained soldiers move easily from one aspect of such a dual relationship to the other as one of the professional skills of military life.

The third level of authority bearing down on junior division, and most omnipresent from morning to night, were the under officers, appointed from the senior division and selected for their combination of zeal, a sense of discipline, plus general probity and promise as future officers. In each company a senior under officer and three junior under officers took care of minor disciplinary offences, assisted by officer cadet sergeants and corporals, again drawn from the senior division. Each SUO and JUO had carefully defined though quite considerable powers. Every morning they took the breakfast roll call, a brief parade plus inspection on the square, before breakfast. Junior division came in for the most detailed inspection each morning, the intermediate division the next in severity, and the senior division rather less. Every dog had its day, and in the junior division we accepted that for six months we would suffer, just as we would in turn impose authority on our juniors when we moved up to senior division.

Standards of turn-out and drill were exceptionally high. Two or three years later I heard that some of the 'bull' had been relaxed, perhaps because the Army Council was becoming alarmed at the number of suicides at Sandhurst. There were two during my time there, and the academy did its best to forget the matter after informing parents. The method of suicide was usually a very unpleasant one. An officer cadet would somehow contrive to bring a live round back from the rifle range and blow his brains out.

Foot drill was taken by the permanent staff, who told us in matter-of-fact terms that the standard required was the highest in the British army and therefore the highest in the world. We had no means of judging this, but it was very exhausting for several months. Extra drill was laid on for those who lagged behind or fell foul of authority. This was a fairly gruesome affair at dawn, where the junior under officer on duty could vent his feelings at having to get out of bed early to bark commands on the square at sunrise. There were some refined pieces of minor torture, including one which began with the command to stand to attention, then to lean forward from the waist in order to 'lay down arms' by numbers. This was promptly combined with the order 'Quick march', so that you waddled forward, horizontal from the waist, clutching your rifle near the ground. At a cracking pace at dawn, this item of unorthodox drill was agonising.

During the regular drill parades by day the atmosphere was gentlemanly but curt. All members of the permanent staff addressed us corporately as 'gentlemen' and individually as 'sir'. We in turn addressed them as 'staff'. The staff sergeants knew that their prime duty was to get the junior division in their company up to scratch in time for the series of drill competitions which would set one company and college against another, with pitiless exposure, on the academy square. The judges of the competition were senior drill instructors from the Guards training depots at Caterham or Pirbright. Our Sandhurst drill sergeants stood to lose face both in their own mess and before their fellow guardsmen from the depots if they did not aim high. Drill was therefore very sweaty indeed, and probably one of the factors contributing to the cases of suicide.

Nevertheless our drill instructors had been selected carefully. They were not sadists, nor were they stupid. They were proud men; perfectionists, who knew that they were the best drill instructors in the army, yet shrewd enough to realise that a little light relief is necessary now and then in order to get the extra effort which might carry off the coveted title of 'Sovereign's Company'. This was awarded to the company whose general all round excellence placed it above all the other companies at Sandhurst, and gave its members special privileges within the academy community.

Our own CSM was a witty, well educated man in his mid-thirties, which was young by the standards of other CSMs. We liked him and so it was not too surprising that at the end of our junior term we won the Sovereign's Company competition.

The earls, viscounts, barons, and other peers of the realm among our ranks produced some unusual exchanges on the college square, and CSMs and RSMs probably enjoyed them. If a viscount was idle on parade, he was addressed with due regard to title.

'Viscount Melgund, sir, that was an idle Order Arms, understand, sir?'

'Staff!' the viscount yelled in acknowledgement.

Sandhurst was still a repository for what remained of the British aristocracy. This changed over the next two decades, and lists of Sandhurst entrants would include fewer and fewer of the names which belonged to a British Almanack de Gotha. To read some of the names on the lists in the small grey book supplied to us in 1949 was to encounter the last refuge of British aristocracy in its dispersal and retreat. Among us were the Lord Wraxall; the Master of Napier; Viscount Melgund; Baron Nugent; Cortez-Leigh; Adams-Acton; Bowes-Lyon; Burke-Gaffney; Cavendish; Champion de Crespigny; Dracopoli; Gun-Cunninghame; Gregor-Macgregor; Sebag-Montefiore; De Sales la Terriere.

It would be easy, and quite wrong, to say that Sandhurst at this time was a snob's paradise. Certainly there was a bond between those who were at Eton together, or Winchester, Wellington, or Rugby. They wore better clothes than the grammar-school boys, and some of them went off to debutante parties in London at the weekends. But in the harsh, spartan life of Sandhurst, one was impressed by their courtesy, their self-effacement or self-depreciation, and by their consideration for others. In this they gave lessons in tolerance and good manners to the more rugged boys from the grammar schools, just as they probably gained something in turn from the individuality, the self-reliance and perhaps the forthrightness of those who had been spared some of the disadvantages of boarding schools. There were no separate camps or unofficial cliques: the daily round was too demanding for that to be possible.

We were told that our conduct at all times must be that of 'officers and gentlemen'. The terms were meant to be complementary. A great many devices were built into our formal instruction as well as the conventions of life in the academy in order to bring this home to us. We dined formally each evening, and though post-war austerity made dinner a khaki battledress affair, blue patrols were later brought back at Sandhurst, so that we matched the officers who dined at high table.

The officers joined us at our tables once a week for 'Band Night'. These were extended occasions, the tables groaning beneath heavy pieces of silver and candelabra, the Sandhurst orchestra fluting away at the end of the hall until the Loyal Toast was drunk. Band Night was one of the earliest Sandhurst traditions, though some of our officers confided that it used to be rather different before the war, when gentleman cadets (as they were then known) threw bread rolls at the band if they disliked particular tunes.

The waiters at table were all retired soldiers, many of them veterans of 1918, some with waxed moustaches, all with straight backs. The same veterans cleaned our rooms and did minor polishing of our kit. Like Oxford scouts, they were shrewd in their dealings with the young men they served and knew precisely what could reasonably be required of them, or when one of their gentlemen needed to be taken down a peg. Mistreating servants was the cardinal sin at Sandhurst. Offenders quickly learned that if the academy had to choose between a servant and any one cadet, the cadet was expendable. It showed a good sense of priorities.

We each received a copy of the Royal Military Academy Standing Orders to guide us through the conventions, formal and informal. I still possess my copy, and it makes odd reading today, but one can see how the various parts made up the whole for a military academy seeking to combine self-discipline with gentlemanly conduct. In the booklet, several paragraphs are devoted to 'paying of compliments' to cover various contingencies when in uniform and when in 'plain clothes'. (The term 'Mufti', we learned, was non-U in the regular army.) We were required to wear some sort of headgear at all times, unless in sports wear. When an officer cadet in plain clothes met an officer or one of the civilian lecturers, standing orders stipulated: 'He will remove his hat in a gentlemanlike manner.' If hatless: 'He will turn his head and eyes to the right or left.'

Bicycles were the common means of getting about from one lecture to another during the day, and the rules for paying compliments when cycling were equally specific for the officer cadet.

'*In Uniform:* He will not salute. He will ride to attention [*sic*], keeping his head and eyes to the front and both hands on the handlebars.'

'*In Plain Clothes:* He will ride to attention. If wearing a hat, he will remove it in a gentlemanlike manner.'

These directives produced curious scenes in the academy grounds. An officer cadet cycling along would suddenly jerk bolt upright, his arms and hands rigid, his back straight, the bicycle wobbling dangerously whilst the passing officer returned the salute or compliment, or civilian lecturers gave slightly embarrassed waves. Once or twice, when the bicycles were thick in both directions, perhaps three or four abreast (though this was against standing orders), a collision would occur and the officer or lecturer inadvertently causing the mêlée would try to ignore the tangle.

Minor rules relating to dress and turn-out on bicycles extended even to cycle clips: 'Trousers will not be tucked into socks', a standing order stated. For those who wore spectacles, 'Only army type spectacles will be worn on parade.' These were the round, steel-framed spectacles which the hippies of the sixties were to adopt from the military, along with badges and uniforms.

Hair: Officer cadets will keep their hair cut in a short and soldierly manner. If a moustache is worn no portion of the upper lip will be shaved. [Hitlerite or Lupino Lane variants were not allowed. Handlebar moustaches were equally forbidden.]

It is an offence for Officer Cadets to travel as pillion passengers on motor cycles.

Officer cadets will not give presents to any member of the permanent staff. . .

Mail: Officer Cadets will request their correspondents to address mail to: Initials and Name- *Esq* (*not* 'Officer Cadet . . .').

Tradesmen: Officer Cadets will not interview tradesmen within the RMA Sandhurst, except the tradesmen appointed to the RMA Sandhurst, ie Tailor, Barber, Saddler.

Visitors: a) Officer Cadets may not bring male Other Rank visitors into the grounds of RMA Sandhurst if they are in uniform. [A curious, ambiguous rule, this. One wondered at the propriety conferred by plain clothes.] b) Officer Cadets may take Lady guests to the following places *only* within the grounds of the RMA Sandhurst . . .

There followed a detailed list, composed mostly of 'ante-rooms' and 'Main Entrances' but the list of permitted spots included 'Gymnasium (For visit only)'; and 'Library (For visit only)'. These rules provoked ribald comment among ourselves.

Overall, the standing orders aimed at a gentlemanly celibacy combined with devotion to duty. All forms of temptation were ruled out, from the giving of presents to consorting with males or females in the academy precincts 'except for visits only'. Official rates for paying our servants were tightly controlled.

> The Dormitory servants will be responsible for and are paid for by the Crown for cleaning Officer Cadets' rooms, buff equipment and boots. Any Officer Cadet who desires further service may pay his servant a maximum of 3/6d per week for cleaning buttons, belts, web equipment and shoes, and for brushing clothes. Such payments are entirely voluntary and are authorised for extra service. The only exception to this rule is that Officer Cadets requiring servants to clean hunting boots and riding clothes may pay up to an additional 1/- per week.

The rates of gratuity have presumably been revised since 1949.

The curriculum was tightly organised and we were expected to get from one parade to the next in a matter of minutes, even if this involved a quick change from drill order into physical training shorts and vests, or from grubby denims at motor vehicle maintenance into battledress for drill, or a sand model exercise in the 'model room', where tactics and strategy were discussed with miniature replicas of arms and equipment.

The tight timetable produced general mental alertness, and we soon adjusted. It also brought quick thinking, and trained us towards a general economy of physical movement which would last us for life. This ability to execute movement of any sort efficiently, precisely, often with great speed but also calmly was just one of the less obvious skills secured for us. Advanced weapon drill with a variety of small arms weapons added to this training until it became second nature. At the same time, physical abilities were promoted by sophisticated assault courses, including various acrobatic hazards. From a platform in a tall tree, a trapeze wire about fifty yards long descended across a small lake. You took hold of a small pulley with two canvas straps and launched off, rifle, steel helmet and all, clinging to the canvas straps as you hurtled down the wire across the lake. In one or two cases a man's nerve gave out, and he dropped about twenty feet or so to the mud of the shallow lake, to be picked up promptly by PT instructors in fisherman's waders.

Another test, which few of us found agreeable, involved crawling through a series of narrow underground tunnels, hardly more than three feet in diameter, which were beneath the soil and totally dark. Some of the tunnels ended in cul de sacs, and you had to back up awkwardly, trying not to panic. We followed each other down the dark entrance holes like moles, and as we crawled in the blackness most experienced the primitive, fleeting fear of claustrophobia. One or two genuinely panicked when they came to a dead end in the black tunnels. We heard later that the test had been drastically modified, with light wells let down into the blackness, after some unpleasant incidents of near terror.

CHAPTER 11

War and the Humanities

Military instruction in the lecture rooms and on sand model tables made us familiar with the organisation of all arms of the services and their functions and interrelations in times of war. The roles of the army, air force and navy, the tactics of a divisional attack, of the waterborne and the airborne landing in the different phases of war, from attack and defence to consolidation or withdrawal, were all examined with great thoroughness.

We began with the principles of war. In order of priority (though it was stressed that to negelect any one of them was to risk military disaster in the field) the nine principles were: maintenance of the object; maintenance of morale; co-operation; concentration and economy of force; security; surprise; offensive action; flexibility and mobility; good administration.

At first it seemed odd to find administration in lowest priority at number nine, but much later, in battle conditions, one saw the need to give primacy to other principles. We devoted two sessions to discussing and analysing the basis of morale. Not surprisingly we concluded that good leadership was the central clue to good morale. This produced another discussion on leadership, a theme which occurred and re-occurred throughout our courses.

Other forms of instruction—including lectures on the great military commanders of the past—were used to strengthen the lessons on what makes for good military leadership. One of the best lectures was delivered by a tall, spindly major in the Guards, exquisitely spoken, who began a lecture on Napoleon thus:

'Napoleon was a brilliant general. He was also extremely small, and is said to have possessed only one testicle; a fact which, if true, would seem to throw doubt on the axiom: "The bigger the balls, the better the man." ' When the audience subsided, a very good lecture followed.

An occasional piece of ribaldry was not frowned on, provided we recovered quickly and composed ourselves. On rare occasions the military instructors played tricks on each other, no doubt after a good evening in the mess. A lecture by a much decorated veteran of Alamein on defence in desert warfare was always accompanied by his constant admonition, 'Never sit back and wait for the enemy: hit back at him all the time.' To illustrate this vital lesson on the art of defence, and to leave it firmly in our mind, a visual aid in the form of a large printed scroll was unfurled at the end of the lesson. When he had concluded his address, the major walked across to the scroll and, with his usual panache, flicked the tied string with his cane. The scroll unfurled crisply enough, and the cryptic lesson was revealed. The bold black lettering should have read: 'DON'T SIT: HIT!' But some hidden hand had added an 'S' to the last word. The major took it in good part: 'Another good lesson,' he said, tapping the scroll. 'Before the battle, always have a good shit: it concentrates the mind.'

Apart from straightforward military instruction, academic studies up to the first or second-year undergraduate level were taught by the gowned civilian lecturers. All of them were university graduates, though heterogeneous in manner, appearance and intellectual power. Military history was compulsory, together with mathematics, but we were allowed a considerable element of choice in our courses in the humanities, languages and the sciences. I opted for government, economics, French language and literature, and the modern novel.

The instruction was generally of a high order, with some exceptions. Our mathematics teacher was a large, lazy man who closed his eyes for an extraordinary length of time as he expounded the calculus or advanced trigonometry. On one occasion, for a bet, a member of the class stood on a seat and cavorted briefly during one of the closed-eye disquisitions.

Economics took us little beyond marginal analysis, price theory, and the theory of the firm. Government studies were well taught, with analysis of various applications of the Westminster model, explorations of socialist theory and international relations in their historical contexts.

There was little incentive to go out in the evening. We were allowed a late pass on request, but a reason had to be stipulated. The attractions of Camberley were few. A few pubs, a cinema, and one or two restaurants just outside the gates for officer cadets who had appetites beyond the Sandhurst fare. Our food was adequate but plain. We had a well stocked shop, called the Fancy Goods Store, or FGS, and a bar-cum-cafeteria next to it. All were contained within Old College, and similarly in New College and Victory College. After dinner, we would gravitate there to discuss the day's doings, tell anecdotes about our lecturers and their courses, or gossip about the officers' wives who turned up in cars to collect their husbands now and then, and who invited us to tea or sherry occasionally in the early Victorian houses just inside the Sandhurst walls, standing like a row of galleons marooned in a wood.

There was some talk of sex, or the absence of it at Sandhurst, but obscenities were not part of the cult. Possibly the acute physical demands of our training acted like a bromide, leaving little energy for sexual fantasy. Homosexuality was discreet and rare, considering the cloistered existence. Opportunities for heterosexual encounters were almost non-existent inside the academy during the week. One of the waitresses in our cafeteria was thought to be available if approached discreetly, but she was about thirty, and not to everyone's taste. One of my friends at Sandhurst did make the approach and formed a liaison during his senior term. She was divorced, had a small house just outside the academy gates, and he would visit her regularly, returning before dawn by climbing the wall where, he said, a succession of night visitors had made the rough way smooth.

For a brief period two attractive young girls appeared as waitresses in the café in New College, but this could not last. Excessive numbers of Sandhurst men began taking their morning and after-dinner coffee in the New College café. Instead of two hundred or so, numbers rose to about five hundred, the queues were vast, and the system quickly broke down. Moreover, it was rumoured both waitresses became pregnant within a matter of weeks, and they vanished promptly. The authorities having made a bad tactical error, the girls were replaced by middle-aged ladies.

Another tactical error was the importation of a very attractive physiotherapist to the medical centre. Her job was to deal with injured or strained muscles after sporting fixtures. The bush tele-graph did its work, and the graph of injuries shot up. Scores of officer cadets turned up for treatment, their damaged tendons mostly in the region of the groin. But the therapist was coolly professional, and word came back that she dealt savagely with the male erection by means of a karate-style finger flick, bringing instant detumescence. Injuries on the sporting field declined.

Sex was thus extra-mural, and tended to be sought in London on weekends. The richer officer cadets had cars garaged in Cam-berley and went off at weekends to dance at night clubs. They exchanged opinions about favourite night spots as we read our newspapers in the ante-room before dinner. The 'Four Hundred' in Leicester Square seemed to be one favourite, the 'Chanticleer' another. Night clubs were so much beyond my pocket that the thought of going there did not even occur to me.

Ostentation of any sort was officially discouraged, and certainly frowned on among ourselves. But here and there a rich officer cadet might sport his wealth by the time he reached senior divi-sion and was heading for an expensive cavalry regiment or the Brigade of Guards. One rich man in New College announced that he could no longer tolerate the food there, and each evening a Rolls Royce brought a waiter from some restaurant, carrying a huge tray with silver chafing dishes. This went on for two or three weeks until the officers discovered it and he was ordered to desist.

As always, military life alternated between harshness and humour; between spartan vigour and a concern for professional standards in everything we did. The concern with standards was seen most of all at a weekly drill parade, on Saturday mornings, when the whole Royal Military Academy paraded together before the adjutant. It was not a simple matter to form up the three colleges, each composed of senior, intermediate, and junior divisions, a total of nearly a thousand officer cadets, plus scores of drill sergeants, twelve CSMs, and dozens of assistant drill instructors. All this on one drill square. It was an immense task to weld together our foot drill, rifle drill, saluting, marching and counter-marching, with or without academy band. It was RSM Lord's job to bring this to near-perfection.

Once the three colleges were assembled on parade, RSM Lord took over, with an impressive display of command and a hawk's eye for an offending rifle butt an inch out of line somewhere along—or behind—the front rank. After a good deal of shouting, with company sergeant majors leaping about and chivvying their men, the sharp staccato of boots stamping on gravel, RSM Lord finally judged himself satisfied. Then he would pivot about, pace stick under his arm, march up to the adjutant, Major Earle of the Grenadier Guards, enthroned regally on his big white horse, and report for all to hear across the huge square:

'Royal Military Academy Sandhurst, present, and ready for your inspection, SIR!' The last word literally echoed around the buildings behind us.

RSM Lord was required to hand over to the adjutant precisely at the stroke of eleven hundred hours. The early preliminaries —beginning with platoons and companies scattered around the three colleges—required the utmost precision of timing. On one Saturday morning the process of getting a thousand men into line, each line as straight as a builder's rule, took longer than usual, and the clock above the square showed the hour only a minute off eleven o'clock. But RSM Lord insisted on perfection in the lines. Then it was 'Stand Still' as he marched up to salute the adjutant. Major Earle, on his big horse, was seen by a thousand officer cadets to look pointedly at the academy clock on the central tower and then at his watch. Then he called out, for all to hear:

'One minute late, Mr Lord. It won't happen again.'

'SIR!' RSM Lord barked.

Everyone had heard, but no damage was done to reputations. The two men respected each other. If military precision is the aim, then those who fall short of it must acknowledge it, so that it is put right next time. There were no excuses. It was one of the rules of the game.

Sometimes we marched across to Old College Square to practise for the passing out parade of the senior division, who ended the parade by a slow march up the academy steps to Auld Lang Syne. At these rehearsals the adjutant also practised the Sandhurst spectacle of urging his horse up the steps into the stone-flagged hall of Old College. The adjutant's horse was normally a placid beast, though we gathered that before these parades the groom was authorised to feed it some sort of bromide or tranquillising fodder. Even so, the final stages of the ceremonial up the stone steps and into Old College were never quite so smooth as the distant onlookers, or, later, the television cameras suggested. In order to close the parade smoothly, the adjutant needed to follow fairly close behind the senior division as they slow-marched up the steps and into Old College. But whilst there was plenty of room on the square, the hall of Old College was very crowded at this moment. The adjutant's horse, already nervous from negotiating the stone steps, was likely to react nervously inside the echoing hall, and we would hear Major Earle (his sword drawn, as these state occasions required) bellowing very loudly, 'Out of the bloody way, out of the bloody way.' Then he would dismount, and the groom would take over, holding the big white beast in this improbable location until the square was finally cleared of all cadets, visitors and onlookers.

The adjutant would also dismount and hand over his horse to the groom before inspecting us on the normal Saturday parade. One morning the horse was very fractious, and the groom had a difficult time calming it. We watched silently from our ranks as the groom stroked and patted, holding the reins firmly. The assembled academy stood to attention waiting for the adjutant's inspection. Then the unthinkable occurred. The big horse kicked its rear legs smartly in the air, reared up, shook off the groom, put down its head and galloped straight towards the front rank of the Royal Military Academy. As the animal shot across the square straight for us, military stoicism underwent its severest test. To break ranks was a heinous offence. But if the animal continued its bolt it would plough through the three ranks like a locomotive at speed. I was a little to the side of the direct line of attack in this mad scene. Just as a tremor developed swiftly along our ranks, the animal saw the obstacle and veered off, hugging our front rank like a fence. Out of the corner of my eye I detected a rippling effect along the front rank. The unhappy groom was sprinting along after the beast. Both horse and man were pursued by the adjutant's curses.

'Somebody stop that bloody horse,' he shouted several times. Again out of the corner of the eye I could see a brace of CSMs waving their pace sticks like conductor's batons, but prudently side-stepping at the last moment as the animal galloped towards the inviting green of the cricket pitch. Once there, the beast settled down quietly to munch the outfield.

We recovered our composure as the adjutant continued his inspection, all of us straight-faced. But as the inspecting party passed along, I thought I saw a bright, strange light in RSM Lord's eyes and the slight twitch of a smile.

Occasionally we were inspected by the commandant, General Sir Hugh Stockwell. He lived in the big house inside the grounds beyond one of the cricket pitches. He was liked as commandant, partly because he brought a touch of eccentricity, even a fugitive swashbuckling air from his previous job as commander of the Sixth Airborne Division in the Middle East. He had a fondness for wearing cavalry-twill trousers with service dress tunic, and woollen mittens on cold days. Although a big Daimler car was at his disposal he preferred to go to his office on a miniature collapsible motor bicycle of the type dropped by parachute for airborne landings. Squads on drill often had to give an 'eyes right' as the commandant hurtled across Old College parade ground on his tiny scooter. When he spoke to us during inspections the gravel voice was friendly, with a joke usually in the offing. Generals like 'Hughie' Stockwell are a uniquely British product. We may or may not produce them in the technological armies of today or tomorrow.

H

CHAPTER 12

Mens Sana

The crowded timetable at Sandhurst gave little time for independent study or intellectual inquiry. Hardly an hour of any working
day was free of formal instruction of one kind or another. Games,
or some sort of sporting activity, were more or less compulsory, on
an honour system by which the individual had to settle with his
conscience if he dodged the column. I swam, because I enjoyed
it, almost daily, and was in the Sandhurst team as a swimmer
and water-polo player. At other times I threw the javelin, pole
vaulted and played soccer.

Like all of my contemporaries I had little or no time to rumin-
ate on personal happiness or contentment, though I had already
firmly decided that I would not remain in the army, but leave
at the end of my five-year engagement, or sooner if possible.
Voluntary withdrawal from Sandhurst was a difficult and expensive matter. One or two did so, but only after protracted
discussions and the payment of some sum which corresponded to
the amount the Crown was deemed to have spent on us. One
gathered that the army claimed to be spending more than two
thousand pounds of the taxpayers' money on us before completion of the Sandhurst course, and buying yourself out of the
army was supposed to be commensurate with that. No precise
sum was specified, but as I was without private means I decided
I would simply have to be patient.

A certain amount of genuine intellectual inquiry took place
in the Sandhurst clubs and societies which advertised meetings
on the notice boards outside our dining hall. One society was
devoted to military history, and papers were read on the strategy
and tactics of famous battles in the past, though not on the politics and diplomacy behind the battles. There were no political

clubs. Politics, especially party politics, were frowned on. This seemed a strong convention rather than an express prohibition— derived no doubt from the dislike military men often feel for party politicians.

Apart from military music on parade and the academy band, which metamorphosed into an orchestra on Band Night, music hardly featured at Sandhurst. As a music lover I had to rely on the radio, turned well down to comply with regulations, or concerts in London on occasional weekends. But the gap was partly filled when I discovered a collection of gramophone records, nominally belonging to a music society at Sandhurst which apparently had ceased to function. The collection had originally been brought together during the war, perhaps by charitable gift, when Sandhurst was a convalescent centre for war wounded. I also unearthed a turntable and a gargantuan loudspeaker. The whole contraption was much too large to fit into my room, but I had permission to settle in a corner of what later became the Indian army museum, a cavernous hall filled with cases of uniforms, medals and mementoes of the North West Frontier. The high ceiling and these various unorthodox sound absorbers made the acoustics excellent. Upper registers of the Brandenburg concertos achieved a special resonance as in the nave of a cathedral. There was plenty of Bach in the collection, some Mozart, Brahms and Beethoven, and choral music from early polyphonic and baroque to some of the modern composers. I was rarely disturbed as I sat at the end of the hall, hidden behind glass cases full of sabres and cutlasses. Once or twice one of the big double doors at the end of the hall would open tentatively and a puzzled face would appear. But military institutions inflict on everyone the constant fear of trespassing on the preserves or privacy of superior officers, so I capitalised on this and listened in seclusion.

Sandhurst insisted on some form of sport or outdoor activity for everyone, and I discharged this on the soccer field. During the season I began to spend more time at the game, profiting from the coaching of Walley Barnes, the Wales and Arsenal captain, who came each week to show us some of the finer points. The combination of his patient coaching and good humour improved our game remarkably. I became a regular inside forward for Sandhurst and travelled a good deal with the team.

In the senior division I became secretary of the academy soccer club and so had to arrange fixtures with other secretaries, some of them at Oxford and Cambridge colleges. I enjoyed our visits to Oxbridge, though the colleges were not always certain that eleven men would turn out, and much variety of dress materialised on the field. A game with New College began with only nine men opposing us, dressed attractively in an assortment of white sweaters, or borrowed jerseys, and thick shorts designed for squash courts or fives. Their play was a curious amalgam of grammar school thrust and Winchester langour. After ten or fifteen minutes of play the remaining members of the New College eleven drifted on to the field, with genial apologies for having overlooked the college notice board.

As the 1940s gave out, the faces at Oxford and Cambridge were changing. Some war veterans lingered, bearded and bulky, with desert boots and a loaf of bread at lectures to take home to their wives and families in caravans parked outside Oxford. But this post-war breed was now being displaced by younger national servicemen who made up part of the soccer team, with here and there a recent sixth former, shy as the ale loosened tongues and bawdy military jokes flew about the panelled rooms after our games on college fields.

We also played the medical students at the London hospitals: St Thomas's, the Middlesex, 'Bart's'. Their stock of jokes over beer reached sophisticated levels of obscenity. And there were annual matches against Cranwell and Greenwich, where inter-service rivalry carried over into beer-drinking competitions after the game. The tribal rites always ended in song, with beer mugs slammed down with gusto on tables to stress lewd rhymes and refrains.

Walley Barnes improved our game continuously, and I began to play for Corinthian Casuals, mostly older men drawn from the professions in London, who played attractive, zestful soccer on their ground at the Kennington Oval.

The journeys to Oxford, Cambridge and London reminded me that my future was not with the army, though at the same time they brought welcome diversions from the daily grind at Sandhurst. The private world I relied on, and into which I silently retired—sometimes during lectures or model room discussions—continued. This private world embraced an inclination towards the arts which, I supposed, was partly derived from a Celtic strain which found its expression in a love of music and the things of the senses.

I also experienced another struggle, common enough at that age for those who walk under the night stars and think about life and death. For a brief moment the struggle surfaced and I showed my hand. I turned to the senior resident chaplain, Colonel Battersby, to ask if I could come and talk, as I was considering taking the Catholic faith. He was a reserved man, but his honesty and integrity were so palpable as to be almost intimidating. After dinner on a blustery autumn evening I went to the house sheltering under trees behind the Sandhurst chapel. He sat opposite me by the fire in his study. I sat silently for several seconds, looking at the fire. Then he said simply, 'You'd best talk first.'

I told him I was thinking of accepting the Roman Catholic faith, and thought I would discuss my reasons with him. I mentioned some of my fundamental doubts, and the unanswered questions which brought me to his study. I think he understood quickly enough that I was seeking certainties; something to give order to an untidy universe; something to put away my suspicion that the universe was indifferent; something to avoid the thoroughgoing determinism towards which I was moving.

I stayed for an hour or so. He was patient, restrained, honest. He offered no certainties. There was faith and there was reason. He did not attempt to cement them together, or argue that the two might ultimately be two sides of a coin. I returned to my room and put the confused stirrings into an annexe of the mind, where they remained subdued, if not ignored.

CHAPTER 13

Kindred Spirit

On alternate Sundays we were required to attend chapel, though it was left to the conscience of the individual to honour this. In junior division attendance was probably one hundred per cent. In senior division devotions lapsed. An occasional ukase from the authorities bolstered attendance for a few weeks. When not due for chapel we were allowed a thirty-six-hour pass away from Sandhurst. Many went home, but as it was too far to go to Northumberland on a short weekend I went to London or stayed the weekend at the homes of Sandhurst contemporaries. In London we could stay at the Bernhard-Baron Club for officers and officer cadets, near Gloucester Road.

Sandhurst men had a familiar style of dress in plain clothes, which marked them out at Waterloo Station when the Camberley train debouched. In wet weather, a belted riding mac; a soft trilby hat (usually from Herbert Johnson in New Bond Street, but possibly from James Lock in St James's Street if the man was trying for the Guards or the Cavalry after Sandhurst); a suit, and highly polished brogues. In dry weather, a blazer with bright buttons, cavalry-twill trousers, brogues, and again the trilby hat. There was also an unmistakable Sandhurst walk—crisp, putting the heels down firmly, the back straight and the body moving forward in one plane, the right arm swinging in time with the feet, the head up and the chin held back. Ex-Guardees might have the faint suggestion of a swagger at the hips, but to accentuate this was to convey Other Ranks status.

London was still bomb-scarred, but the age of austerity was giving way to greater freedom and the ending of controls. Small restaurants along the King's Road gave a good meal with wine for ten shillings. There were concerts at the Albert Hall and opera at Covent Garden. In a basement at 100 Oxford Street, Humphrey Lyttelton and his band played New Orleans jazz twice a week. On Saturday nights it was crowded out, a social mix of young East-Enders, debutantes, students from the Chelsea Polytechnic, from other art colleges, and from RADA. I discovered the place by accident one Saturday when I heard Dixieland jazz somewhere beneath my feet and traced it to a long, low basement below a restaurant. Lyttelton's band was probably at its best at the time, playing in a free-ranging New Orleans style. Wally Fawkes was on the clarinet and Keith Christie on trombone. The upright piano was stripped of its covers and a tuba player turned up now and then. The vocalist was George Melly, a younger, thinner version of the jowled critic of the sixties.

Young people danced, some in existentialist blue-and-white striped jerseys brought back from the *caves* in Paris. It was easy to make friends. Most came regularly and some remain in the memory for their striking appearance. Lucy Forbes-Robertson, gypsy-like, with hair down to her waist. And a tall, Rupert Brooke figure with a handsome, pagan look. He wore bedraggled foulard bow ties and flesh-coloured shirts beneath a crumpled tweed jacket. At least they looked like flesh-coloured shirts, until he came near and you saw that he did not wear a shirt, merely a bow tie supporting a grubby loose collar. Everyone knew his name—Alexander Plunket-Greene. His girl friend was strikingly beautiful and they were a familiar sight in Chelsea, hand in hand. He was a wild young Irish aristocrat given to playing the solo trumpet along the King's Road and getting arrested. He lived in Markham Square in a setting of rumpled disorder and unwashed wine glasses, where parties began suddenly, escalated alarmingly, and where his neighbours, including the Sitwells, complained of the goings on. A few years later he married Mary Quant and changed the face of King's Road.

Chelsea was still a community. The speculators had not yet moved in and poets and painters came into the pubs near closing time from studios near the Thames. Augustus John held court at the Cross Keys. In the anonymity of the capital, eschewing Sandhurst dress, I sowed wild oats. One affair was platonic and it

outlasted all the others. There was an air of Bloomsbury in the
tweed cape she had wrapped about her. She was nineteen and
had recently come back from the Sorbonne. What did she do
now? Nothing really, except write poems and not send them in
anywhere. It was not a sensual encounter, but a chord had struck.
Her name was Josephine. Her parents were divorced. The family
had literary cousins but her mother had tried to put her through
finishing school and a debutante season. The Sorbonne had been
her escape, a flight from the horrors of 'coming out'.

We wrote to each other and her letters flowed. She told me of
the books and poems she had been reading and she wrote out
the passages which moved her. Our letters grew longer. At Sandhurst the military life became easier to bear, inhabiting a less
essential corner of the mind. Looking at her letters now—the few
which survive in a box of bric-à-brac—they bring to mind what
adults easily forget: the passionate intensity with which young
people seek the answers to fundamental riddles about their own
lives—and which direction they should take—as well as wider
questions about the human predicament. The scrawled letters
also recall how voraciously we read. In one exchange we were
enthusing about Panait Istrati, a little known Rumanian novel-
ist, son of a Greek smuggler, who married twice, settled in Nice,
then moved on. He wrote painfully in French and died in Paris
in 1935 of tuberculosis, believing in absolutely nothing. His
novels *The Bitter Orange Tree* and *Balkan Tavern* seemed to us
to sum up the twentieth century much more completely than
Eliot's *Waste Land*.

In another letter she wrote of Gunnar Bjorling

. . . who I've discovered in one of the poetry quarterlies. I will write
down the poem that I love here. You will understand what is meant by
it and no doubt have experienced the same feelings yourself. I call it
'the flash of eternity' but whatever one may call it, it is always there,
ready to be felt, only we hardly ever feel it, depending again on what
degree of 'spirituality'—for want of a better word—we possess, for I
think, don't you, that the only 'true' life is essentially spiritual, and
our physical one mainly the resulting complement of an inward drive
towards fulfilment of spirit—if you see what I mean. Anyway, if you
are still there, here, *enfin*, is part of the Bjorling poem, translated by
Marjorie Gerhardi:

Shadows in the Sand

Quiet summer among the trees
And by the shore, without stars,
Without wave or wind
And all is
All for a moment
All that was and will become . . .
I am as near the light
As is the light of flowers in
The dark lands of the graves.

It is not the woods, the waters,
Not graves nor earth nor death;
It is the silences of smiling people,
It is the sorrow of the old upon all shores.
It is the dirge of the unconsolable,
   —the last watch together,
With the bloom of life in the
Months of the young dead
That the waves swept gently.

In other letters we enthused about Garcia Lorca in the Spender translation. We discovered Sidney Keyes together, devoured all of Dylan Thomas before he became a cult, and cast around for texts or commentaries which, we felt, encapsulated our beliefs. We were convinced we knew what was wrong with the world; that we could put it to rights. Seditious stuff for Sandhurst, but by now I had learned to compartmentalise my mind and activities. As this friendship grew to an intimacy of mind we came to depend on each other. She provided the escape from the military environment which I needed, whilst at the same time she needed a close friend as she struggled to confront the issue of her relationship with her mother, who commuted between a house in Montpelier Walk and another in the Hebrides on the island of Eigg. We felt that we had each found a special kindred spirit, so much so that we addressed each other and signed our letters with the abbreviation 'KS', from that day to this. The friendship survived through her short, disastrous marriage with a painter, through the time when she almost died of pneumonia, then her mother's death and then to Hampstead, where she lived in Belsize Park for a time with six cats, named respectively Beulah, Clytemnestra, Stanley Matthews, Nice Person, Freddie Ayer, and

Macavity-Mau-Mau-MacTavish. Then she settled down with fewer cats, a huge Saint Bernard dog, a minah bird, and Mohan, in a house by Christ Church Hill. At times we walk on the Heath and think back.

Her mother was a tall, spindly woman of the type one sees in Harrods instructing a salesman to send a small brown loaf by van. We got on better than I expected. Perhaps she felt that if her daughter was taking someone to bed, or was got with child, then better a Sandhurst man than a Chelsea poet: one could always write to the commandant. Eccentricity seemed to run in the family. She addressed Josephine as 'Spinster' and was addressed in turn as 'Crone'. Neither seemed to mind. But divorce had brought loneliness, possibly shame among her circle of friends, and she needed her daughter's company on the island.

She spent part of each year at Eigg island, and commanded her daughter, 'You must bring him to Eigg. You need company.' To Eigg I went on the next vacation. The boatman dropped me on the shore just below Kildonan, the converted farmhouse on the sea turf above the rocks. KS helped us to unload the cartons of stores Crone had ordered from Mallaig. Crone appeared correctly dressed in tweed suit and brogues.

We were left to ourselves. We walked on deserted beaches and wrote poems, reading them to each other before we went to bed. One or two were published later when we bothered to send them in. Crone went to bed early, probably to drink herself to sleep in the four-poster bed. We sat up, listening to Beethoven. The wind came off the shore, down the chimney, scattering sparks from the logs in the open hearth. We drank coffee from fluted Dresden cups, blue and orange and gold, and read more poetry. Sandhurst seemed a universe away from the dark islands. Coll, Tiree, Benbecula; the Celtic grail; the titan chords; Jura, Canna, Skye: the names were poetry enough.

At the end of the week I went back to Mallaig. The gannets were diving and gulls looped where she sat on the sea turf, her hands clasped about her knees. Now and then we waved to each other, until she was a distant speck on the island shore.

CHAPTER 14

Senior Division

In the senior division at Sandhurst the few privileges to be earned came our way. Some of our contemporaries had gone back to the ranks, unable to keep up with the pace of work. One had killed himself in his room. Two or three were RTU'd for stealing personal belongings from friends' rooms. We were surprised when we heard the names—they seemed pleasant, inoffensive persons, but their actions may have sprung from some subconscious wish to get expelled, to get away from the strain.

In military instruction we had passed on from the organisation of the Division and the Armoured Regiment to higher things: the Corps, the Army Group, and military strategy. We also learned to drive, and were supposed to have a detailed knowledge of the internal combustion engine, but I had some mental block about this, and could never grasp how a carburettor or a clutch worked. We drove three tonners around Camberley and Aldershot with instructors by our side and learned how to get a motorcycle out of deep mud on Bagshot Heath. There were visits to tank training grounds at Bodmin Moor and Lulworth Cove.

Towards the end of our training there was an ambitious five-day tactical exercise on Salisbury Plain. We slept in the open, more or less maintained ourselves under battle conditions, and fought a night-time battle in battalion strength. On a fine May night we marched forward on Salisbury Plain, under brilliant flares, with live ammunition cracking above our heads, the machine guns firing 'fixed lines' from rigid clamps mounted on scaffolding to our front. Trip wires set off minor explosive charges a few yards ahead of us, simulating minefields as we moved into 'enemy' territory. Thunderflashes were tossed at our feet to test our nerves as we prepared to charge with fixed bayonets. In the sky above, pink magnesium flares turned night into day, and now and then green and red Very lights signalled different stages of the battle. The whole thing brought home the incredible noisiness of a full-scale battle.

As usual with army manoeuvres and battle exercises, we were utterly exhausted, but oddly pleased with ourselves—as though we had come through some testing initiation rite. We finished our five days on Salisbury Plain with a sense of elation. Possibly the long night marches, the deep silence of the nights on sentry watch or on patrols under a full moon, worked some primitive alchemy. We were out of the caves, hunting in packs, near Stonehenge, our senses of sound, sight and touch—even smell—alerted and attuned, giving them an interdependence not normally required in the safe streets or houses of civilised life. Back at Sandhurst we resumed the rule of law.

The daily training programme was still strenuous, the time-table crowded every day. Mornings were given over to military instruction, both model-room discussions on strategy and tactics, and various forms of outdoor instruction, from weapon drill and live firing to assault courses, driving instruction, and cross-country exercises. Afternoons were for sports, and early evenings included the more strictly academic side of the teaching programme, when plain clothes were allowed.

In our senior term we studied military law, so that we would know how to act as defending officer in a court martial. The various offences against military law and the Army Act produced flights of invention in our off-duty hours, including improvised court-room testimony relating to charges of sodomy. Among ourselves the mimicry portrayed the average NCO giving evidence, the satire in a sing-song delivery:

> 'Sir, on the fourth of the fourth I was orderly corporal and was proceeding in the direction of the cook-house after "Lights Out" when I heard the unmistakable chink of a coin of the realm. On investigating closer, I beheld the h'accused, bending over boxes, ammunition, with his trousers and drawers cellular down on his ankles. His posterior was h'exposed and at the high port.
>
> 'I also beheld the other h'accused party, Private Ramrod, placing himself close to the h'accused's posterior in a position which indicated they was about to commit sodomy. I approached both parties to the

act and in a manner of speaking I said to them "And what is this 'ere?"
Both parties attempted to desist from their action but had difficulty in
standing to attention owing to circumstances. I placed both under open
arrest and reported to the orderly officer. . .'

As the day for our 'Passing Out Parade' approached, we were
asked to state the regiment or corps we would prefer to join when
commissioned. Most men asked for the regiment they had joined
before coming to Sandhurst, but some were now attracted by the
technical work in the Royal Engineers, Royal Signals, or the
Gunners.

The fact that you nominated a particular regiment or corps
was no guarantee that you would be 'gazetted' to it. There were
many more applicants for some elite regiments than there were
places, and word came to us that some regiments had so far ex-
ceeded their quota with previous Sandhurst intakes that there
would be no vacancies whatever for our class. This proved to be
the case with the Black Watch, which had been heavily over-
subscribed by every post-war Sandhurst intake. When the company commander told me this, it came as a blow. Now I had to
run my eye down the Regiments of Foot, Highland, Lowland,
and the Border Country.

Not having served in any other Scottish regiment, I had no
claim on them, and the few vacancies were already over-subscribed
by native born Scots. The Border Regiment seemed the next
best thing, as its historical connections lay with Cumberland,
Westmorland and my own native Northumberland on the English side of the Scottish border. I put my name down, though
again this was no guarantee of getting one's first choice. We were
told, as crisply as a military academy knows how, that many who
nominated regiments of the line would find themselves commissioned to the Royal Artillery, the Royal Engineers, the Royal
Army Ordnance Corps, or worst of all, the Royal Army Service
Corps. There would be no appeal, for you had no rights in the
matter. Getting your first choice depended on your placing in the
final order of merit—a forthcoming event where, on the evening
before Passing Out Parade, the senior division assembled in a
large lecture hall to hear the list read out—beginning, as in

beauty competitions, with the last on the list in order of merit. If your name was read out very early, your chances of gaining your preferred choice of regiment were practically nil. It would mean you were bound for the lowly RASC—'Al Sloper's Cavalry', as we called it, for some derogatory reason.

I was commissioned to the Border Regiment and with the others received a tightly furled paper scroll, in copperplate, signed by King George VI. We were given a week's leave before joining regimental depots to await further posting orders. The depot of the Border Regiment was Carlisle Castle, an impressive bastion just beyond the cathedral, with a broad moat, a portcullis, massive walls, and the crenellations of the picture book castle. When I arrived at Carlisle, the regiment was abroad, at Mogadiscu in Somaliland, but due to return soon on completion of a three-year tour there.

The mess was a comfortable place, and four or five officers on depot duty lived well in an atmosphere of polished antiques, open hearths and leather padded settles, drinking their beer from silver tankards as they warmed their bottoms at the blaze. I was gazetted to the regiment together with a Sandhurst contemporary, Kenneth Hodkins, who was killed in Korea little more than a year later. A regiment of the line had a special interest in its latest Sandhurst additions to the mess, and in due course we were 'dined in' to the regiment. But at the depot, dinner was a leisurely affair. We dressed in dark blue number one dress, then gathered in the ante-room over sherry or a beer until the mess corporal came in and clicked his heels to the major commanding the depot, when we ambled through to the dining room.

Conversation in the mess soon brought diminishing returns. There seemed to be a strong convention that subalterns tittered respectfully at the senior officers' jokes or anecdotes. The basis of many of the jokes was one or other aspect of the class system: incompetent butlers and the gaucherie of people with ideas above their station were common themes. As a relief from the tedium I signed out to meals at the weekend and walked on the Roman Wall, exploring the western sections. I already had a fairly detailed knowledge of the eastern portions from my Northumbrian boyhood.

As we had no obvious duties at Carlisle Castle, Kenneth Hodgkins and I were temporarily attached to the East Lancashire Regiment, stationed a few miles away at Hadrian's Camp. We were to remain there until the Border Regiment arrived back in England in two months' time. This gave us our first taste of regimental life as subalterns. The East Lancashire Regiment had suffered badly from a series of home postings in the United Kingdom, and many of the officers were seeking secondments elsewhere, especially abroad, in better climates. Morale was low, and discipline in the camp was slack. My chief memory is of one national service officer who had developed the art of dodging duties—or 'skiving', as the army put it—to a consummate degree. He was going up to Oxford that year, and each day hid himself in an empty office out of sight at the end of a long wooden balcony beyond the company office. He took the precaution of taking along a sheaf of official looking notes and pamphlets and the always impressive clipboard, to which he could turn in the unlikely event that he was disturbed. Once installed, he gave his whole day to reading thrillers. One day I walked along the balcony to find him with his feet on the trestle table casually reading a paper-back thriller. Intrigued by his casualness, I wondered what he would have done if the adjutant or some senior officer had walked in on him. For answer, he pointed languidly to the window, which had been carefully opened on its steel strut exactly 45 degrees so that it gave a mirror reflection of whoever approached down the long wooden balcony.

Our company commander was a grey-faced, pessimistic character who drank late and alone in the mess each night and could usually be found asleep in his blue patrols in the ante-room at breakfast time each morning. Sometimes he tottered down to company lines at mid-morning and gave out one or two token jobs to each of his subalterns, including our resident 'skiver'. He would accept the instructions glumly and, once out of sight of the company office, he would take off his service cap and kick it hard into the air, like a Rugby player aiming for the touch line. He

made no secret of his dislike of the army, but his idleness probably did not begin with National Service. If it did, it carried over
into his Oxford career. A few years later, at Oxford, I was discussing with some friends the idlest person we have ever met,
and his name cropped up again. He had left a vivid impression
on some of my contemporaries at Oxford. Too idle to look for
digs in his second and third year, he was reduced to living on a
punt moored near Folly Bridge. He kept his one crumpled suit
on the back of a door in a scout's kitchen, and borrowed a towel
about once a term to take a bath. Now and then he bundled off
his shirt and underclothes to the laundry. The tale was that he
knew the time had arrived to send off his laundry when he flung
his socks at the wall and they stuck there. Perhaps Oxford wit
had adorned the tale. He ploughed and went off to plant tea in
India or Ceylon, ignorant of the twilight of Empire.

Fortunately, the tedium of Hadrian's Camp was broken by a War
Office posting to the Small Arms School at Hythe in Kent. This
was to be an eight-week course in basic weaponry for platoon
commanders. There were about thirty of us on the course, and
when we gathered in the ante-room at the mess we met Sandhurst
contemporaries, now transformed in regimental dress.

The aim of the course was to make us highly proficient in the
handling of every basic weapon of the infantry platoon. These
included the .303 rifle, the Bren gun, the Sten, the pistol, the
PIAT anti-tank weapon, the two-inch mortar, and the grenade.
The instructors were extraordinarily skilful and the weapons were
like toys in their hands. Their accuracy on the firing range was
deadly and effortless. And whereas we had thought ourselves perfectly capable with the weapons up to this moment, we now saw
the difference between the ham-fisted amateur and the really professional weapons expert.

After endless practice we could lie on the floor with our rifles
and press the trigger without toppling the 12-sided threepenny
piece balanced edgeways on the foresight. We could also operate
the bolt with a smooth action, getting off ten rounds in twenty
seconds. On the firing range we learned to strip and re-assemble a

jammed Bren gun smoothly and swiftly. Our marksmanship was brought nearer to Bisley standards by subtle tricks of the marksman's trade. Indeed, the best among us went off to Bisley that year and one or two won prizes. At the pistol range we learned how to hold the weapon firmly and steadily—discovering also how intensely difficult it is to fire accurately with a pistol. Before long we could hit the man-size targets at the lethal points.

The course was gruelling, the daily pressure unrelenting. Thrown against each other in the usual competitive atmosphere, and knowing that the commandant of the Small Arms School would be sending a confidential report on our work and progress to our regimental commanding officers at the end of the course, each of us felt the strain as the intensity of the course increased. One incident served to illustrate the strain.

It occurred after lunch one day. Our schedule that morning had been intensely exhausting on the firing ranges by the sea. For four hours we doubled along the slippery shingle with guns and spare magazines, knees wobbling as the feet slid away. The morning ended with an assault course, each of us carrying weapons over high stockades and swinging from ropes across minor chasms, the guns slamming into the hips, leaving huge bruises. One or two collapsed temporarily at the end of the assault course, gasping for breath. But when lunchtime arrived, we rallied and stumbled into our quarters to change from mud-spattered denims into plain clothes, as this was Wednesday half-day.

After lunch we relaxed with coffee in the ante-room, immersed in the papers, one or two falling into a light doze. I went back to my room to write letters, but what happened in the somnolent ante-room after lunch was described graphically next day. The door was thrown open suddenly and one of the subalterns strode in, wild-eyed and dishevelled, his arms flailing theatrically. He shouted out, 'I can't take any more of it, I just can't!' Then he drew from his pockets two hand grenades, tugged out the pins and flung the grenades on to the carpet.

As the grenades rolled across the rug, the fuses popping in an authentic fashion, a remarkable scene occurred in the ante-room. Within seconds subalterns were bunched at the door, pummel-

I

ling and shoving hard to get out. A man sitting by the French window burst it with his shoulder. Another sitting by the empty hearth scrambled up the chimney. On the sofa, a nervous ensign in the Guards fainted.

It was a convincing hoax. A few minutes later, when they realised that the grenades had been dummies, those in flight returned to find the hoaxer quietly reading a newspaper. Most were angry, but he was a big, athletic man, and after a day or so most agreed that it had been a pleasant diversion from the daily grind. At the end of the course our weapons-handling had improved several times over.

CHAPTER 15

Initiation Rites

The Border Regiment was now on its way home, to be stationed at Barnard Castle, on the border of Yorkshire and Durham. I was posted there with Ken Hodgkins to join the advance party now preparing the camp for the battalion.

We were at Deerbolt Camp just across the river from Barnard Castle, spread about the crown of a rising knoll. A few officers forming the advance party were already there, plus a score or so of soldiers and warrant officers, moving about the empty billets and rooms, checking inventories, fittings and equipment so that the outgoing quartermaster could sign everything over to the Border Regiment quartermaster when the job was done.

The officers' mess was not yet organised, with only the minimum of furnishings, indifferent food, and the beginnings of a bar. The regimental silver would arrive later, so we made do with crude cutlery and saucers of salt when we dined. The temporary commanding officer was a major with a fondness for administration. He gave us various jobs to do each day, which mostly involved peering into the more remote corners of the camp. Two or three of the other subalterns were Sandhurst men, and told us frankly that the battalion was not looking forward to a home posting, even though this was long overdue. Apart from this, the commanding officer of the First Battalion was about to retire from the army, and no one yet knew who the new CO would be, so there was a period of uncertainty.

The present CO, Colonel Cooper—or 'Coop' as he was known to everyone—was genuinely admired and respected by the regular members of the battalion at every level. His whole career had been given to service with the regiment, in one or other of the battalions, and he knew or had known almost everyone living who had served with them, besides a great many now dead. With his retirement due, there were some difficult days ahead for the First Battalion, as events were to prove.

When they arrived, with their crated luggage and equipment spilling desert sand and dust, their hats bleached by the African sun, their skins deeply tanned, they looked out of place under the grey skies of early spring in Yorkshire. But no one had much time to think about this as company lines were taken over, the main cook-house opened up, and buglers began to sound the various calls of the day from Reveille to Lights Out. The mess silver was uncrated, and dining became more civilised. As new Sandhurst subalterns, we were 'dined in' to the regiment on the first guest night. This was intended to be both a courtesy and a compliment to the new members of the mess; but over the years it had acquired the additional function of initiation rite. Both Ken Hodgkins and I knew we were to be tested as well as dined in.

To begin with, our drinks for the evening would be charged to the other members of the mess. We knew that danger lurked here, and picked our way carefully as the sherry decanter cruised above our glasses before dinner.

'Come on, you're not drinking that,' a senior major called out at my elbow. 'Straight down with it. Go on. Let's have that decanter over here. Don't they teach you to hold your wine at Sandhurst these days? Come along there...' He guffawed in a friendly fashion.

As the bronzed faces circled about, perhaps remembering their own initiation, one or two grins suggested that other things were in store. At dinner, the wine waiter was clearly under instructions to see that our glasses were never empty, and by the time the port had been passed three or four times I was beginning to feel the effects. After dessert we went back to the ante-room, and tankards of beer were thrust into our hands. The senior officers stood near the fire, warming their backsides as the subalterns carried out their unofficial duties of plying us with beer. We politely declined a contest with the senior subaltern for downing a pint in one gulp, knowing that the resulting disaster would be much worse than the chorus of jeers from the hearth. I was still steady, as I had taken the precaution of eating a big meal, and hoped that the ballast below the belt was now soaking up the beer. But the rites were not over. The adjutant, a keen rugger player, announced 'Mess Games', and the armchairs were pushed back to the wall, leaving an arena of bare rug.

Various tests of endurance followed. One involved the heavier officers carrying lighter members of the mess on their backs, seeking to unhorse the other riders. Patrol jackets were cast off as the jousting continued. The field officers staggered beneath their platoon commanders. The evening's champion emerged, as one rider after another was torn from his mount. Once or twice horse and rider collapsed on the hard floor. I took my turn, my legs gripped by my company commander wheezing beneath me as I manoeuvred to get a purchase on a fellow subaltern. But I was dislodged by a more experienced rider.

Another test was a familiar one: lying full length on the floor, facing opponents, hands clasped, discovering who could force down the other's forearm to the floor. Our mess, like most others, had an undisputed champion, who bared a huge forearm and took the whole thing very seriously.

A final game—which was sufficiently dangerous and strenuous to end the revels—required the whole mess to split into two teams. One team crouched as in leap-frog, but head to tail, forming one long column anchored against a wall. The other team ran up, leaping in turn on to the long 'horse', and clinging there as the others followed behind. The overall test was to see whether the anchored team could support the whole of the other side on its back. Once again, there were clever arts. Peering backwards through his own crotch, the last man making up the 'horse' could waggle his bottom just a fraction as a man sprinted forward and took off, hands forward, legs wide apart. With a well-timed waggle, a real flier might miss, hit the floor, or the wall. But once aboard, the advantage lay with those on top. By redistributing their weight, they could bring unbearable strain to one side of the 'horse', forcing the weakest man to his knees.

We finished up with one bloody nose among us, but as there

were no cracked collar bones this was thought to be a fairly placid evening's sport. More beer was drunk, and the senior officers retired to the sofas. One of the subalterns found a piano in the adjacent room, and we gathered there to sing bawdy songs. The two subalterns being dined in were required by convention to stay to the end, however prolonged the night might be. In the months that followed I was to hear the same bawdy songs sung repeatedly, and though they palled fairly quickly I was impressed by their sheer grotesquery on this first evening.

All regiments, it seems, have (or had) their own bawdy songs, and it may be a pity that sociologists or other students of folk-lore did not record them before they were dispersed by the creeping death of the county regiments in the army reorganisation of the fifties and sixties. For the songs revealed a good deal about the various theatres where the regiment had served, and incorporated snatches of foreign tongues—especially Arabic and Urdu—which added colour to the song, and heightened the bawdy. They mostly recounted the deeds of legendary characters, male or female, who either possessed immense sexual prowess or were comically impotent. Eskimo Nell—of universal acclaim—was of course the prototype of the first.

I did not note down in my journal the songs and refrains the Border Regiment had fashioned as its contribution to the canon, chiefly because I rarely felt in a writing mood after guest night in the mess. There must have been one exception, however, as a rough piece of HMSO stationery shows some scribbled lines in my hand. Deciphering the scribbles, they bring to mind the tuneless fortissimo of those late evenings when the silver tankards swung rhythmically and red-faced Englishmen sang tunelessly.

One favourite song described the activities of three Baghdad homosexuals. It was sung to a Gilbert and Sullivan melody, and was characteristic of the simple, heavy metre favoured for mess songs. Certain key words could be shouted, rather than sung; these words tended to be obscene, and came conveniently at the end of a couplet or a line. The Arabs were hymned thus:

> We're three black bastards from Baghdad,
> We like to live a life of ease,

And we don't give a rap
For the pox or the clap,
For to spread disease is our design.

Refrain

For we've got 'em, yes we've got 'em,
Black spots all over our bottoms,
And we don't give a rap,
For the pox or the clap (etc)

Now we only charge you half a crown,
You can have it standing up or lying down,
And we don't give a rap (etc)

Another song described the deviance of a young Englishman
who, for some reason, had the name of 'Craven, A'. It was sung
to the tune of 'Steamboat Bill, Steaming down the Mississippi...'
and provided a full biographical saga, though my scribblings
supply only a few of the verses:

CRAVEN, A

Now listen all you people, won't you listen, till
We tell you of a pederast from Muswell Hill,
He grew up with his aunties down at Camberwell
But the first three words he said were 'Bloody-Fucking-Hell.'

Refrain

Craven, A, never heard of fornication,
Craven, A, never had his beans,
Craven, A, he went in for masturbation,
Playing with his foreskin in the school latrines.

Now at his public school he caused quite a to-do,
He buggered all the prefects and the masters too,
He got expelled, so the records say,
For putting it up the Duke of York on Founders' Day.

Refrain

Craven, A, never heard of fornication (etc)

Now his entrance into 'Varsity was quite grotesque,
He went and laid his penis on his tutor's desk,
M' tutor said, 'If it drops off at a later date,
Please remind me, and I'll use it as a paperweight.'

Refrain

Craven, A, never heard of fornication (etc)

Others were more earthy, their usual *leitmotif* interminable bouts of copulation. Another snatch in my scribbled notation records exploits at a tavern:

> Now the landlord's daughter Nellie was as nice as could be
> She always brought her c— up with a cup of tea,
> I've been through her so many times, the laws declare
> Her vagina constitutes a legal thoroughfare.

Ken Hodgkins and I managed to get through our first guest night, dined and wined into the regiment, still standing on our feet. We had passed the initiation rites. We could hold our wine; we could play strenuous games after a gargantuan meal; we could hold down pints of beer and sing profanely as our tankards were filled again and again. In short, we were fit to be subalterns in the Thirty-Fourth Regiment of Foot.

'Coop', the retiring commanding officer of the First Battalion, took his leave in a quiet fashion. There was no ceremony, not even in the officers' mess. We heard that he had ordered this himself. Like the old soldiers in the song, he merely faded away. But in fact there was a small presentation in the ante-room. It was early evening, and the CO's car was outside to take him to the station. Only about four or five officers were in the mess as he put his head round the door to say goodbye, simply and without ceremony. The senior subaltern produced the silver chafing dish the officers had bought, with its very simple inscription, giving the years he had commanded the First Battalion. He stayed for a drink, briefly. A shy man, reserved, utterly contained, and master of his feelings, he had given his life to the army, and had probably been passed over for higher command because of his complete devotion to the regiment and the men in it. I think his eyes misted over slightly as he shook hands with each of us—young men beginning the career which he now concluded without ceremony in a barren little brick mess in Yorkshire. Then he was gone. He was not an old man, but he died within a few years. The War Office had by then announced that various county regiments were to amalgamate. The Border Regiment was to lose half of its identity in a new 'King's Own Border Regiment'. He would not have wished to witness the obsequies.

Various names were canvassed as possible COs for the battalion. They had to be officers of the Border Regiment, of course, and some of the more distinguished lieutenant-colonels serving on the General Staff were mooted. A favourite starter was Colonel Svegintzov, a highly intelligent staff officer of Russian extraction, with a distinguished record in Intelligence. But 'Zog', as he was called, was either reluctant or needed for higher things. Finally a new CO was appointed, Colonel Kington-Blair-Oliphant. He was a very tall, thin, but also a very sick man when he arrived, with various ailments gathered from jungles and deserts abroad. His health deteriorated further in the damp Yorkshire weather. After one or two absences in hospital, he left us.

Morale in the battalion suffered further. The 2IC, now acting as CO, was an unpopular man, not of the Border Regiment originally, and therefore doubly unwelcome. Finally we heard that another Border lieutenant-colonel was coming to us. But chance remarks among the senior officers indicated that it was a controversial appointment. One major applied for a posting immediately. Another went off to command a territorial battalion some weeks before the new CO arrived.

The rumour among the subalterns was that he had an ungovernable temper. It was not a happy prospect. He arrived in the mess one teatime, a broadly built, glowering man with huge shoulders. He ignored the subalterns when they handed him sandwiches. We stood about the ante-room, wondering whether we would be introduced. Finally the adjutant repeated our names. He nodded briefly to each of us, his eyes concentrating on our dress and turn-out as he scanned. Then he went out with the adjutant and acting CO to walk round the camp. When he was gone one or two said 'Phew!', but no more than that.

Our new CO was a disciplinarian, but he was probably right to think that the first necessity was to discipline the regiment in order to build its sagging morale. Whether he went about it the right way was open to question. A series of written orders came out of the adjutant's office concerning saluting—including the precise number of steps other ranks would hold their salutes when passing officers. The officers were not spared. In a blunt talk in the mess he told us that his main impression was of slackness everywhere, and he reminded us that there were no bad soldiers, only bad officers. He was right on this point, of course, and perhaps it was also true that several years at a boring station by the sea in Somaliland had made everyone slack.

He said he intended to put things right—and quickly. Every officer would study the history of the regiment, using a brief potted history available from the Regimental Depot. In future, whenever wet weather kept the men in their billets, the junior officers would give them lectures on the history of the regiment, and quiz the men on the details of the more illustrious battle honours. Added to this, every man in the battalion, from CO and adjutant to the office clerks, would spend from noon to 12.30 on every weekday on weapon drill, with rifle, pistol, mortar, or whatever his allotted personal weapon happened to be.

This last instruction led to some bizarre scenes about the camp each day when the bugle sounded the call for weapon practice. The warrant officers and sergeants supervised the men in the billets, whilst officers gathered about the company offices, drew their pistols and snapped them repeatedly at each other. My company commander was an eccentric man who tried to bring actualité to pistol play by inventing a game of hide-and-seek around the company offices. The major would come creeping along the rear wall of the company office and then leap around the corner, snapping his pistol wildly. The daily game became a Western-like *High Noon*, with each of us flattened against one or other wall of the huts, edging along with pistol at the ready. Within weeks, the officers began to ease up on this daily chore and the order faded away gradually for the men also.

The rainy day lectures on regimental history never quite got off the ground. There were limits to the amount of personal embarrassment the subalterns were prepared to risk in front of the national servicemen who made up most of the other ranks. One keen subaltern did attempt it once or twice, but was so aghast at the men's ignorance when question-time came round that he dropped it.

In peacetime soldiering the main task was to keep trained for battle by an admixture of drill, weapon training, occasional exercises, and sport. I was a platoon commander in 'A' Company, and had the good luck to serve with the most interesting of the company commanders, Ambrose Trappes-Lomax. He was something of a throwback, an eighteenth-century character, or even earlier; an arcane figure, educated at Stonyhurst, and given to forgetfulness. He had a slow, deliberate manner, which came partly from a studied punctiliousness but also from a mind that seemed to work very slowly. He groped towards the correct, the fair, the just decision. If any of his sergeants or officers or men were threatened by authority from without, he instinctively took their side until the full facts were laid out—with necessary patience— before him. He never gave offence, such were his qualities of courtesy and charm, and even the new CO had a special affection for him. When daily parades were over, Ambrose would walk around the camp like a squire surveying his demesnes, clad in a heather-coloured, hairy tweed suit with plus fours, a gold hunter at his waistcoat, brogues, a tall shepherd's crook, and a tweed cap, which he doffed elaborately when returning salutes.

He carried his vagueness into the company office, with a trusting nature besides. One day he solemnly took off his hat in the company office and turned to the CSM.

'Sar'n Major, please tell me, do I require a haircut?' He turned round obligingly. The CSM looked at the silken undergrowth about the neck and ears.

'It's just about due, sir,' he said, tactfully.

'Then I shall have one today,' Ambrose announced.

For some reason he preferred a hairdresser beyond Darlington, about twenty miles away, and he went off that afternoon. He returned without a haircut, but driving a gigantic black Buick car as long as a tank, more suited to a Chicago gangster syndicate. Somehow he had been diverted en route, and a car dealer had sold him this monster, an old model which gobbled up petrol at a rate which even Ambrose soon found he couldn't afford. But meanwhile, and before he sold it at an immense loss, he added another picturesque note to the camp, peering out at the level of the bonnet of the car in his squire's cap, or edging the monster gingerly round the officers' quarters like an ocean liner.

The other officers were a composite mixture, such as one could find in any regiment of the line. All of them placed a high value on personal honour, integrity and the call of duty. They were available to their warrant officers and men for twenty-four hours of each day, and if this required them to cancel a social engagement, or even a planned holiday, there was no question of priorities: duty came first, pleasure was always secondary.

As the months passed, their other qualities became observable. They were men of limited imagination, conservative in their habits, but by no means stupid. For professional tasks they could think quickly, but above all calmly, especially when it was most necessary—under severe pressure or when the battle raged. They never panicked: to 'flap' was the deadliest sin of all. Nothing was calculated to lose you more face than to 'flap'. They were men who, having given their word, could be relied upon completely to keep it. This was partly the code of the English gentleman, but partly a necessary by-product of their professional job. If a leader does not keep his word to the men he leads in peace or war, defeat is the first possibility.

As officers they were the product of a particular society in a particular historical phase. They adopted the verbal cadences and some of the mannerisms of the traditional upper class, which helped them to give orders without self-consciousness. And given the purely technical requirement of authority and a chain of command for war, it was better that they gave orders confidently rather than self-consciously.

The brigade commander was a remarkable man who quickly rose to higher rank in the army. Brigadier C. P. Jones was the most efficient man I have ever met. A small, neat, witty man, he would gather all the officers in his brigade, senior and junior, into a large model room in Deerbolt Camp to discuss tactics. For this, he had laid out on the floor hundreds of toy-size models of every vehicle, gun and wireless network in the Infantry Division, as well as its supporting arms, from Armoured Regiment to Artillery, Royal Engineers, RAOC, RASC, ambulance brigades, etc. With the officers gathered above him on terraced seats, Brigadier Jones would first identify the main units, using a long pole, and then get down to detailed discussions, ending with questions. Throughout, he showed an astonishing mastery of the intricate relationships between the thousands of items of minor hardware, and hundreds of small units making up the division. A typical problem was presented thus.

Question: In a vital battle, the main radio link between divisional headquarters and medium artillery to the rear is cut through air attack, then three other radio links are severed: what alternative links can be used *without* risking loss of contact between other vital forces in the division?

For answer, the brigadier would take his long pole and trace half a dozen labyrinthine wireless connections, forwards and backwards, laterally, diagonally, transforming the model into a huge intricate chessboard of a thousand pieces, with the grand master demonstrating the best moves, and a few unorthodox alternatives for good measure.

He did this wittily and with casual ease. He knew the name and face of every officer in his brigade, so he could toss the question about the three regiments present in the model room, setting them against each other. It was a masterly demonstration of professional competence, entirely at odds with the familiar caricature of the blimpish brigadier.

A few weeks later, the discussions became even more intricate when the brigadier added air support with both fighters and bombers to the strategic and tactical exercises. For this, he was joined by another remarkable man, John Barraclough, commanding an air station at Middleton St George nearby, and soon to rise to air vice-marshal. They made an unusual pair—Jones small, neat, precise; Barraclough very tall, with a leonine head, relaxing at question time with an immense Churchillian cigar. At the end of the general discussion the two commanders took up some

of the finer points of air-ground liaison, both of them showing
an extraordinary grasp of the intricacies of the separate commands. Once again, it became a bravura yet light-hearted performance, given not to impress the onlookers, but arising from
a total absorption in the complex, technical problems thrown up
by the maze of vehicles, tanks, guns, wireless stations, communi-
cations and supply lines. Overall, it was an astonishing display of
complete professionalism.

The brigadier was also good company at dinner, ready to tell
a story against himself, and on one occasion, with a touch of
impishness, against the GOC. Apparently the general was in
command of manoeuvres on Salisbury Plain in the depths of the
English winter. Anxious to show solicitude for his troops, like
Henry V before Agincourt, he ordered a jeep to take him up near
the front positions at the dead of night. He wandered quietly
about the trenches, with only his ADC and the local commander
for company, giving an encouraging word here and there as the
intense cold bit into the soldiers on guard. The men were surprised and flattered to be visited by the small party of officers,
especially when they discovered that the GOC himself was paying them a call. Except for one lonely soldier. He had suffered
the sort of day which occasionally hits any soldier in the field.
He had been ordered to dig a trench in the frozen ground here;
then the position was re-sited; then this order was counter-
manded again when another senior officer inspected the positions.
Cold, hungry, exhausted, almost beyond words in his feelings, he
had hardly dug in again when the GOC came forward softly in
the blackness of the night with his words of cheer. 'Now then,
my man, what's the general situation; anything to your front?'

The answer came back bitterly in the black night:

'Fuck all. And you can fuck off for a start.'

The GOC stole away, knowing when to leave the British soldier
well alone.

A good deal of regimental life was made up of afternoon sport,
and as a soccer player I was put in charge of the game. I had
settled to the position of left half in what was then a game of

positional play, quite unlike the fluid permutations of 4, 3, 3 which arrived in the late fifties and sixties. I played at Catterick once or twice for Northern Command against Southern Command. The teams were largely made up of professional soccer players, many of them signed up by leading league clubs, which they would rejoin at the end of their National Service. In some games for Northern Command, our team included John Charles, the Leeds and Welsh international player. He was a trooper in the Twelfth Lancers, a young giant deceptively fast for his bulk, brilliant in the air or on the ground. When our forwards failed to score in the second half of the game, he moved forward from centre half and casually rose above the defenders to flick goals into the net like bullets from a gun.

I travelled about, playing for Army Crusaders once or twice, and this provided some escape from the tedium of life in barracks. Regimental life was proving to be a dreary, repetitive round, without novelty, except the very minor diversions provided when some of the men had to be retrieved from the civil powers after drunken brawls. We had our share of old lags—men who had served in the army for more than twenty years or so, and had known no other life. They had served in every sort of theatre throughout the world, in peace and war, and no doubt they felt very confined in the sleepy little Yorkshire village with its law-abiding pubs and a dearth of resident whores. Some of the men went farther afield to Middlesbrough or Darlington for their pleasures, and when summer arrived there was a sudden rash of camp followers near the barracks. I drove out of the camp early one Saturday afternoon, taking a short cut past an open meadow, to find a flock of whores in summer frocks who had settled there like butterflies. As the driver stopped at a minor crossroad they ran forward to the landrover and there was a brief flurry as we beat them off.

I was orderly officer next day, and at mid-morning I received a message from the orderly sergeant. Two women had been found in one of the billets at reveille and were now being held in the guardroom. I went down, and they were brought out—sorry-looking creatures, pale, undernourished, poverty stricken. The sergeant said they had been found in a billet that morning, brought here by the orderly corporal of the day. The women looked terrified as though wondering whether an officer had authority to have them shot at dawn. There did not seem much point in asking them why they had trespassed on military property; the answer was obvious, and I had no wish to hand them over to the police for trespass, loitering, or prostitution. The best thing seemed to be a warning about the crime of being found on military premises without authority. A guard escorted them to the main gate. When they were gone, the orderly sergeant said he had taken the precaution of looking inside their handbags in case there was any War Department property there. But he only found fourteen separate shillings in each bag.

'Fourteen separate shillings, each handbag?' I asked.

'Yes sir,' the sergeant replied crisply. 'Fourteen men in the billet. What you might call a bob a nob, sir. . .'

For the officers, social life outside the barracks was restricted largely to cocktail parties, given by each regiment in turn, or occasionally by the brigadier. The other two regiments in the brigade were the East Surrey and the Buffs. We dined at each other's messes on regimental guest nights, and extended the repertoire of lewd songs over beer in the late evening. At cocktail parties on Sunday mornings we met the wives of the married captains and majors, some of them attractive in the harassed, ravaged fashion that comes from constant travel, finding new schools for the children, and forever adjusting to new married quarters and new neighbours. They tended to pick up military jargon from their husbands and the environment, which gave their conversation a masculine tone, laced with the milder swear words. One or two mixed minor blandishments with their remarks to younger officers when their husbands were out of earshot.

For off-duty pleasures, Barnard Castle initially offered very little by way of diversion except for the remarkable Bowes Museum standing at the edge of the little town. Housed in a replica of a French château—the detail faithful, but the brown-coloured stone and sharp northern light not quite right to set it off—the collection was and is an extraordinary tribute to the taste and the industry of the mid-Victorian collector John Bowes and his wife, Josephine. I slipped away from camp on quiet afternoons when there were no parades. The rooms mainly contained paintings from the Spanish School, El Grecos and Goyas, as well as French and Italian works by Boucher and Courbet, Tiepolo and Sassetta.

Some of the tedium was relieved as personal friendships developed. In brigade headquarters Geoffrey Sarah, a major in the Royal Army Education Corps, seemed even more out of place in the army than I now felt myself to be. A quiet, self-effacing, scholarly man with a law degree, he had stayed on in the army after 1945, perhaps for lack of that self-interest which makes for early success at the Bar. He was a serious student of music, and his German wife sang with a beautiful voice. They had a grand piano at their cottage near Barnard Castle, and there were musical evenings, with one or two friends invited to play a trio or quartet. Then Geoffrey would accompany his wife at the piano as she sang Lieder, by the firelight, with candelabra on the piano. These evenings provided a refuge from the barracks as autumn drew on.

Some feudal traces survived in the small township, though Barnard Castle had no resident squire. A nearby village provided a substitute in the form of a mountainous lady whose Rolls Royce processed through the main street now and then, filled with Pekinese dogs who yelped and yapped around the chauffeur. The lady did her errands by shouting from the Rolls Royce to the shopkeepers, who hurried out to take her order. She never disembarked, though she was a keen huntswoman and rode with the Zetland, side-saddle by all accounts; but it was difficult to visualise all that bulk poised above a horse.

The military brought no special problems to the township, so far as one could tell, beyond an occasional brawl in a bar. If the local bobby could not handle it, the military police were on the scene very quickly. The officers tended to avoid the pubs, which were considered the preserve of the warrant officers and men. Instead, they would take a drink at the main hotel in the town,

K

or down at the Morritt Arms near Rokeby Hall beyond Greta Bridge.

Peacetime soldiering in Britain is a fairly dreary business, and most professional soldiers prefer postings abroad, provided they do not impose marital burdens or complicate the education of children. We passed the time as well as we could, carrying out training programmes, company and battalion exercises on the Yorkshire moors, and occasionally brigade exercises.

As the months went by, one learned the conventions of the officers' mess. If you were ensconced in an armchair you stood up when the CO entered the ante-room and waited for him to motion you down. You paid your mess bills promptly if you were a subaltern: captains and majors were allowed some days, weeks, or occasionally months of grace if they were overdrawn at the bank or had spendthrift wives. You did not talk politics in the mess, and ladies' names could be mentioned only with circumspection. Subalterns broke the rule when there were no field officers present, but there was an old-fashioned public-school reticence where personalities were concerned. There were other, unwritten rules. I discovered that it was not done to leave a book overnight in the ante-room, nor even in the billiard room, in preference to the limited fare spread out on the sidetable: *The Tatler, The Field, Country Life, Illustrated London News*. I left a copy of Verlaine's poems in an inconspicuous corner of the mess one Sunday night, inscribed by the friend who had sent it as a gift. The PMS (President of the Mess Committee) handed it to me on the Monday with a terse comment.

'Don't leave your property in the mess. It's not done.'

CHAPTER 16

Ulster to Japan

Winter came. It was 1950, the Korean war was six months old and casualties were mounting. The British Twenty-Ninth Infantry Brigade had been badly hit, and I learned that several Sandhurst contemporaries in the Argylls and in the Middlesex Regiment were now dead. Two of them had been killed standing on a hill trying to indicate to American planes that they were bombing the wrong positions. In the January 1951 battles the British forces, now enlarged by the Gloucester Regiment and the Royal Ulster Rifles, were again badly mauled as the United Nations forces retreated south from the Imjin River to Seoul. The newspapers told of cases of frostbite. Other, informal military channels of information said that the incidence had been much worse than the official communiqués admitted.

I had a break from the round of regimental duty on a tactics course at the School of Infantry in Warminster. We stamped around, ice-cold steel on our mittened hands as we clambered about Centurion tanks, or watched the dismal figures of the permanent staff at the school crawling on grey mud under barbed wire for our benefit, whilst staff officers described the tactics through loud hailers. We watched a mock tank battle, the combatants churning up the mud in the valley below like monsters in some prehistoric fight for survival. At the end of the course, we did our stuff in denims, charging through the shattered remains of Imber, the ghost village on Salisbury Plain, with young staff officers snapping at our heels. I was glad to leave the place, with its bad food and its mock-Georgian, Hore-Belisha architecture.

By late April the news from Korea was worse. The Chinese and North Korean offensive across the Imjin River had decimated the Gloucesters and the Royal Ulster Rifles. Only a handful of

officers in each regiment remained, the rest killed or captured. At
Deerbolt Camp the adjutant circulated a note saying that volun-
teers were urgently needed to fly out to Korea immediately. I
thought it over. The Border Regiment was posted for UK service
for another two years on present plans. I had already made up
my mind that I would be resigning my commission and leaving
the army at the end of my five-year engagement. As for the
Korean war, I was not moved by any particular feelings of patriotism, nor was I convinced that President Truman's original
despatch of US troops to Korea—on a weekend when the UN was
not in session—was unmixed with *realpolitik* in the Far East, an
American sphere of interest. Despite all this, however, I was interested, even intrigued to know what war was really like. I wanted
to know what my personal reactions would be; how I would
behave. And Korea was in the Orient. It meant travel, different
horizons, climates, people.

I discussed it with my Sandhurst contemporary Kenneth Hodgkins. He said he would be putting down his name. I could see
that, for him, professional experience was everything. He longed
to put into practice the tactical skills he had learned. We passed
word to the adjutant's office that we would be prepared to go.
Two or three others volunteered, and after a week we learned
who would be going. Kenneth Hodgkins heard first, and within
a day or so he had left for Northern Ireland to join a contingent
flying out to the Royal Ulster Rifles. I heard next. The adjutant
mentioned it with what seemed deliberate casualness outside the
mess at teatime. The full casualty figures for the Gloucesters and
Royal Ulster Rifles had just been published in the newspapers.
The two regiments had been in the most forward positions at a
bend of the Imjin River, where the full force of the Chinese
attack had struck. The Gloucesters had earned the fame—deserv-
edly—but the RUR had been very badly mauled. I walked back
to my quarters, feeling a curious sense of exhilaration, as though
life had been simplified very suddenly. I would either return from
Korea or I would not: it was as simple as that. Life itself gained
an unusual clarity. Even the trees and the gently waving grass
in the fields beyond the camp stood out in sharp relief. I wrote to

tell my parents, stressing that the British troops in Korea would no doubt be given a rest from the front-line positions now. This proved to be wrong; there were not sufficient troops to allow it.

Like Kenneth Hodgkins I was to join the Royal Ulster Rifles, and would fly with the next contingent as soon as I was kitted out at the Ulster Rifles' depot. I was given a few days' leave at home to see to personal affairs and leave behind clothes and dress uniforms, which would not be needed in the foreseeable future. There was no time for sentimental farewells, and perhaps Northumbrian independence gave our family a good basis for this sort of leave-taking. I left the details of a life insurance policy I had taken out, in my parents' name, among my papers in the trunk, and travelled light to Ulster. There would be a delay of at least a week before departing.

The mess at Ballykinler contained an odd company of Irishmen. I wondered whether their eccentricities had made them unsuited to active service in Korea. But someone had to remain behind to look after the shop, and perhaps the remoteness of the camp on this wild shore had contributed to the situation. The senior officer was a major whose first name was Dermot; a large, shaggy man, with eyes like Kitchener's on the poster, and a less organised moustache. A huge hound padded at his heels everywhere he went—inside or outside the mess. Phonetically it was called 'Shaun', an idle, unintelligent beast. One or two of the officers kicked it away from the rug by the open fire when Dermot was out of the room. Dermot's greatest passion was his pack of beagles, which he kept somewhere in the camp. I was forewarned by one of the subalterns that I would be asked—like others before me—to walk the beagles at dawn. The approach came as I was pouring coffee after lunch. I saw Dermot put aside *Country Life* and bring his empty cup across to replenish it.

'D'you like beagling?' He stirred his coffee and fluffed his moustache meditatively. I said that I had never beagled, and didn't really fancy it.

'Well then, perhaps you'd like to walk my beagles tomorrow morning.'

This did not seem to follow from my statement, but logic was not abundant in the mess. As I was about to go to Korea I could afford to risk offence to a field officer. I declined. Dermot avoided me after that, and referred to me in the next few days—when he had to—as 'that officer'.

Dermot was reduced to walking his own beagles, and I gathered they were not the best specimens of their kind. One subaltern who had earlier agreed to accompany the exercise gave vivid descriptions of Dermot lifting his beagles over low fences as dawn came across the Irish Sea.

The 2IC was a worried man called Tom, given to losing documents in his office. Another of the officers, a captain, had got religion and tried discreetly to bring us all to God—especially those on their way to Korea. His sitting room was filled with religious pamphlets, and he had a charming, adept, very Irish way of inveigling you there on one or other pretext, so that the eye could light casually on the literature spread neatly on a green baize card table. He had goodness of soul, and he tried to save any vagrants he met on the roads as he drove about. His trusting nature took some knocks when a man arrived at the camp one Friday afternoon, claiming he was an officer in some battalion posted abroad. The guest was properly entertained throughout the weekend. The man left, well victualled and effusively grateful. When the mess silver was checked certain pieces were found to be missing. The local police knew the rogue, but he was now safely in the Republic, over the border.

The quartermaster's stores at Ballykinler camp held the jungle green uniform and equipment we needed for the Far East. I collected my bush shirts, slacks, shorts, dark green tropical underpants, a dark green towel, socks, a canvas hat and the Ulster Rifles' corbeen bonnet for ordinary regimental wear.

With a pistol in a green webbing holster, two stamped metal identification tabs hanging from my neck—giving name, number and religion—I was ready for war. I parcelled off my plain clothes. Together with the others, I queued at the MO's office for a round of inoculations, which covered everything from cholera, dysentery and beri-beri to a disease named Japanese encephalitis. For this one, the needle went in deep, and one or two in the queue fainted as they came up for their turn. We finally assembled around the transport sheds on a gusty day to be taken to Belfast, where we would take the ferry to Heysham, for London and Heathrow Airport. Before we got away there was the predictable flurry as the whole contingent of officers and men was counted and a discrepancy was discovered. Regularly, it seemed, some of the soldiers tended to vanish across the border as the day of departure arrived.

We were being exported under a secret posting order, referred to as the 'Acanthus' file, portions of which were to be read out to us just before leaving. On this particular day the file could not be found in the office safe and there was a minor panic as the 2IC darted in and out of various doors, the detachment standing at ease, armed to the teeth. Someone blamed the IRA. Dermot had vanished with his hound, having completed his main task, which involved shaking hands with the departing officers, including some who were not departing. The search proved fruitless. Instead of the Acanthus instructions we were each given a small army pamphlet containing advice for those proceeding to 'Eastern Theatres'. I studied this with interest as we crossed the Irish Sea. There was no War Office date of publication, but internal evidence gave some clues. The pamphlet advised taking special care not to wrap puttees too tightly about the legs before the desert sun got up. There were also medical cautions on testing the drinking water for horses. The pamphlet was probably very useful for the cavalry on the North West Frontier at the turn of the century.

We had one night in London, quartered in the Goodge Street Shelter—a disused underground shelter on Tottenham Court Road where we slept in bunks in long lines. We could hardly have a last fling among the London fleshpots as we had only our jungle green to stand up in. But one or two pubs in the area were used to our kind, and some of the old wartime spirit survived as the locals treated us to pints and wished us luck.

There was still some documentation to do next day, and

stragglers were added among the other ranks, until a plane load was complete. We left Tottenham Court Road just as the office workers were on their way home. It was a fine late-spring evening, and as the sunlight slanted across the Great West Road there was a curious sense of detachment and disembodiment as one looked out at the suburban houses and the occupants getting off red buses. As we sat in the chartered bus in our jungle green, with rifles and pistols, green webbing and soft hats, we must have presented an alarming sight at first glance. One or two tired commuters jerked their heads when we stopped at the traffic lights. In the main lounge at Heathrow foreigners started visibly, perhaps certain that guerrillas had taken over and were first securing the airport.

We boarded a BOAC Argonaut—by present standards a slow, four-engined plane. The plane needed several refuelling stops, and in the next two days we stepped on to hot tarmac at Malta, at Nicosia, and at Habbaniya in Iraq, where a wall of heat hit the face as we stepped down in the afternoon. Then Karachi in the middle of the night, where we had small, sweaty fried eggs and limp bacon; then on to Ceylon and a night stop at an RAF base at Negombo. In the early evening we sat on bleached rattan chairs, beneath fans, and were served cool drinks by turbanned servants. The RAF officers wore shorts, starched shirts, neat stockings. Their wives drank gin and tonic and talked of England and rationing, or bantered provocatively until their husbands swept them up from the verandah to go back to their servants and children in the bungalows among the palms.

There were about twelve officers in our detachment, and two dozen men. Half were going out to the 'Glosters', half to the Royal Ulster Rifles. Some were Irish, from the Inniskillings, or the Royal Irish Fusiliers. The remainder were, like myself, from various English regiments. The officers' party decided on a feast in Colombo that evening. The RAF supplied trucks and we drove along the coastal road beyond Colombo, to splash in the sea at Mount Lavinia Hotel. We drank cocktails by the surf, and came back to order dinner at the Galle Face Hotel by the harbour in Colombo. The head waiter rubbed his hands as the waiters spread the white cloth over a large round table. The indoor palms nodded beneath the ceiling fans. We ordered Indian food, and a succession of plates came in, borne by a trail of waiters like bearers on safari. A major who had served in India sent back some of the dishes, ordering hotter curry to be mixed as we belched appreciatively, and downed cold lagers, then reached for another pearled bottle from the pile sweating on a side table. Later, peeling fruit and dipping fingers into scented finger bowls, the major reminisced about the beautiful Anglo-Indian girls he had taken to bed in Rawalpindi before the war. The waiters stood back in the shadows, expressionless when one caught sight of their faces. The head waiter was pleased when he finally counted up the notes we each flung on to the middle of the table. We staggered out to salaams in the fronded hall, relieved to find that the RAF driver was not drunk like ourselves. The truck skimmed along below the trees, and we sang bawdy songs under blue moonlight and scented jacaranda, before falling into our beds, the fan slapping overhead.

We had another night stop at Singapore, and this time the officers divided into two groups, majors and subalterns, the groups now and then meeting each other at the Raffles Club or the Mountbatten Hotel bars. One of the subalterns proposed a taxi for another night spot. The taxi-driver drove us around Singapore's red-light district, suitably named Lavender Street. But we were much less experienced than our corporate bravado indicated, and when the prostitutes tore open the door and tugged at our sleeves, or clambered inside in their floral pyjama suits, we had a sudden fear of the clap or an oriental pox, and pushed them out, urging the driver to get away quickly. We ended up at the Raffles and the Mountbatten again, refuge for the Anglo-Saxons, drinking Tom Collins in the bar among the planters and the businessmen with tuxedos and Home Counties accents.

Our flight ended at Iwakuni in Japan, near the southern tip of Honshu. We were met by British officers in the RASC, and got aboard a requisitioned pleasure boat which took us across the Inland Sea to Kure, our staging post for Korea. Sunlight bounced off the glinting water and tiny tree-crowded islands threw out a delicate beauty, as though our wash would shatter the fragile cameos painted on the water.

The camp at Kure was a big, bustling affair, serving British and Commonwealth troops going to or returning from Korea. Australians, New Zealanders and Canadians mingled with the British officers and men. The Japanese economic miracle had not yet begun, and Kure was a dusty little port, its dock-yards still scarred by bomb damage.

In our camp the military authorities were capitalising on the inferior status of Japanese women. Armies of them swarmed about the camp as servants, waitresses, and laundry girls. I had hardly taken off my jungle green jacket and trousers for a shower when a knock came at the door and a smile came round it. She seized my clothes, even though I tried to show that these were all I had. But she gesticulated furiously and vanished, the Cheshire cat smile still filling the room. I took my shower, shaved and came back wrapped in my towel. My green jacket and trousers were lying across the bed, washed, starched, ironed and hot, the creases knife-edged. My shoes were polished. On the way to dinner I took a wrong turning and found the babble of laundry girls in their hothouse, ironing manically on long tables, chattering like gibbons.

In the mess we were told that platoon commanders—meaning all subalterns—would be going to battle camp for ten days to get 'toughened up' before going across to Korea. Although we might think ourselves fit, the terrain and the conditions over there were rough. We would need to be conditioned. There was an acute shortage of junior officers in the line, so no time would be wasted getting us across. The battle camp was at Hara Mura, eighteen miles away. We would march there with our men in a few days' time. Majors and captains were excused. They looked relieved.

We enjoyed the mess at Kure whilst we could. Japanese girls in white coats waited on us at table, almond-eyed, smiling, silent. In the next day or so we heard about Hara Mura battle camp from the old timers in the mess. We were clearly in for something special. The commandant, by every account, was a madman, given to heavy drinking and occasionally going berserk. He had been up in the hills too long. He was hard on subalterns. He had lost a lot of them in North Africa during the war, sending out patrols on impossible missions. You were glad to get away to Korea after Hara Mura: it was like going off for a vacation.

The march up to the hills was a grinding affair. The permanent staff from the battle camp—barrel-chested men who had performed the march a dozen times—encouraged us on. Occasionally we were allowed a rest, gasping by dirt roads, speechlessly trying to guess how many miles we had covered, how many yet to do. We were told not to drink from our water bottles; it was good training for doing without. Some of the men cheated. The subalterns, temporarily reduced to the ranks, did not care to notice, but tried to set an example. They were anxious not to show any fatigue in front of the men they might have to command in the line within a few weeks.

We stumbled into the camp towards dusk, a ragged, limping crew spun out along the dusty road for a mile or so. We flopped in a clearing under the trees as we were told. The staff strutted about, stripped to the waist, all muscle and rippling tendons. Then the commandant appeared. He waved us down with his walking stick as some struggled to their feet. He also was stripped to the waist. Bare buff seemed to be a sort of special uniform in the steaming camp. He was powerfully built, his walking stick merely an appendage, or more likely his only badge of office about the camp. That, and his brown boots. His denim trousers were sun-bleached and torn. He leaned forward, both fists curled about the walking stick. His black beret was pushed back, exposing his face, moon-shaped, mottled, slightly puffy, sun-tanned.

'Good day to you,' he shouted. The voice was distinctly Irish. 'Welcome to Hara Mura. This is a battle camp. It's not the Ritz. We've got to get you fit. In one week. So it's goin' to be rough, gentlemen. You'll be chased, through the day and through the night, from arse hole to breakfast time.' He let the laughter subside. 'But when we ship you across, you'll be fit, gentlemen; fit to fight and fit to fuck. You'll be soldiers.' What followed was not unlike General Patton's famous ribald farewell to his troops in World War II.

In fact the commandant was a character, and we all grew fond of him. The ingredients of leadership are multifarious and depend a good deal on the particular context for their successful application. This profane lieutenant-colonel was the right man in the right job. He lived by himself in a small bungalow through the trees. Vast quantities of beer and spirits went across there from the officers' mess, but the commandant was always up at reveille, striding about the camp. He looked after the men first and the officers second—as a commandant should. Whenever he came striding through the trees he seemed to build up a buzz of activity ahead of him, like a sound wave before the nose of a supersonic aeroplane. Sometimes it was a scattering, scurrying effect as men dived for cover; usually it was a burst of frenzied application to the job in hand.

There was a system of orderly officer of the day, and when my turn came I went down to the cook-house at breakfast time to look at the food being served. As I approached, I heard a violent commotion going on inside the kitchens. Then two Japanese cooks came sprinting out, holding their heads. Inside the mess hall the commandant was holding a mess tin in one hand and thwacking the table with his stick in the other. One of the soldiers had complained about the size of portions being served. The commandant was bellowing at another Japanese cook, 'Where's the rations corporal? Get me the bloody rations corporal.' He added something in Japanese and the cook ran off. The rations corporal, a beefy, sweating man, came in nervously a minute later. He had obviously jumped hurriedly out of bed. The commandant thrust the mess tin under his nose.

'What the hell's this, corporal? Where's the ration scale? Let's see it. There must be tons of food in this bloody camp. Where the hell's it all going?' And he prodded the corporal ahead of him with his stick. From the kitchens I heard the noise of a stick thwacking the table again, and another Japanese cook sprinted out through the side door. Then the commandant left, muttering to himself, smiting the tables in the mess hall as he strode off. The soldiers grinned and there was a small cheer. I went into the kitchen and found the mess corporal sitting on a chair, breathing

hard. 'That man ain't fit to command,' he said in a shrill cockney. 'He ain't fit to command. He's mad, that's what 'e is, mad.'
But the portions in the mess hall were much bigger after the
incident.

After the mess hall I made my way to the crude NAAFI canteen under the trees not far away. I saw chairs being hurled about
and heard the commandant's stick striking wood, or possibly
human flesh. Once again Japanese servants scuttled out and hid in
the trees. The commandant had discovered that the canteen had
not been swept the night before, or at dawn, and was getting his
message across in his own way. I waited until he was stalking
off through the trees, shaking his head savagely and muttering
to himself.

But the commandant did not neglect the officers so far as it lay
within his limited means. We were quartered in tents, about a
dozen subalterns to a tent. After a day running up and down
the hills firing blank cartridges and digging trenches, we dropped
on our narrow cots for half an hour before going to the crude
cold showers behind the canvas screens. At mid-week, we came
back steaming and exhausted to find a mirage in our tents. A
dozen or so Japanese girls stood about, giggling shyly, ready to
wash our denims and clean our muddy boots. They had already
rigged up some sort of laundry with steaming pots of water. The
commandant had recruited them from the neighbouring villages
and presented them as a surprise. Their duties were primarily to
wash and clean, though some of the subalterns sought further
accommodations in the remaining few days. It worked out at
about one girl for three officers, and most of us had the pleasant
experience of lying on our cots to have our shoulders rubbed with
soothing oil by the giggling peasant girls. One bold man promptly
extended his girl's duties further and sloped off through the trees
each evening to the girl's home, where apparently the parents
welcomed him with old-fashioned Japanese hospitality, and gave
the couple a room. The other girls were not so accommodating,
but these were early days and they were new to the job.

The officers' mess was a crude shack, and four or five Canadian
officers were quartered permanently in rooms just behind the

mess. They had duties connected with radio and telephone com-
munications between Hara Mura and Kure, and with ammunition and other supplies in the camp. They each had a specially
selected Japanese girl, beautiful doll-like creatures, who tended
to all their needs, spiritual and physical. The transient British
officers did not mix with the Canadians, who chose not to eat in
our company. We suspected they had better food. There was a
sort of no-man's land in the little bar at the end of the shack
where the Canadians downed huge quantities of rye whiskey.

Battle camp ended with an extended tactical exercise back to
Kure, sleeping rough for three nights and finishing up with a
dawn attack on a high ridge. The commandant directed the final
operation, and truly came into his own. We rose at three in the
morning and by 0430 were near the base of a hill feature, which
had a few token troops on top provided by the permanent staff
at Hara Mura. The colonel gave us a final exhortation. 'Fire and
Movement' was his tactical credo, and we heard it repeatedly
during the next hour or so. He also had a Japanese battle cry
which sounded like 'Hubba Hubba' and he bellowed this across
the paddy fields as dawn came and we struggled up through the
dew-soaked thickets to the sliding rubble on the crest. In the last
stages of the assault there was a great deal of noise, with a final
glimpse of the commandant aloft on a rock, like Cortez on his
peak, silhouetted against the streaky dawn, slashing the air with
his stick and shouting:
    'Fire and movement! fire and movement! Hubba Hubba.
Come on, come on, Holy Mother of Mary, God dammit Jesus
Christ let's have some fire and movement down there. Attack.
Attack...'
    He was, in many ways, the most unforgettable soldier I met
in the army. A few days later we were shipped across to Korea.

CHAPTER 17

Korea: Imjin River

We crossed to Pusan in an American boat from the naval base at Sasebo at the southern tip of Kyushu. Only five or six of us were going to the British Twenty-Ninth Infantry Brigade, which was composed of the 'Glosters', the RUR, and the Royal Northumberland Fusiliers. The remainder were to form the advance party, either in Pusan or in Seoul, for other British regiments due to arrive as part of the British contribution to a projected Commonwealth Division.

Pusan was a painful sight beyond the waterfront. American trucks drove us through what seemed a shanty town, the dirt roads swirling with black dust. At the railway station the platforms and even the tracks were littered with thousands of refugees, arriving or leaving—it was difficult to say which—as they sat among their bundles without homes or shelter.

A train was waiting to take us to Seoul. Its windows were boarded up, with narrow slits for ventilation. Apparently there was guerrilla activity in the hills between Pusan and Seoul. The train's complement was a mixture of American, Canadian, and British troops. American girls from some voluntary organisation suddenly appeared on the station platform, trim and clean in blue blazers and red skirts, to serve coffee and doughnuts. The first taste of American coffee was strange. We got to know it well in our daily 'C' rations in the months ahead.

The journey was long, hot, uncomfortable. The train had long since lost any proper seats, as its main function was to bring the wounded down from Seoul to the hospital ships in Pusan harbour. We perched on the stark wooden bunks, no more than planks, during the day, and stretched out to sleep as the train thudded through the night. Some played cards and drank scotch.

Others traded jokes across the frontiers of national cultures in the narrow corridors between the bunks, the degree of obscenity increasing as the journey went on.

Seoul had changed hands twice in the last six months, and was now a shell of a town, few of the buildings left whole and again a town of homeless refugees. The dust was brown here, and much thicker, swirling about the vehicles and covering us with a thick layer until face and uniform were dun-coloured, our eyes glinting out like music-hall minstrels'. We separated into regimental detachments and made laconic farewells. 'See you at the sharp end.'

A British truck from the Twenty-Ninth Brigade took us north from Seoul, bumping and swaying between the potholes in the dirt roads. We passed through Uijongbu, which was flattened to the ground, a few thin peasants scratching about in the ruins. Most of the Koreans were heading south, small delicate people who stood aside on the narrow road as the truck lurched past. The fathers carried the family belongings on their 'A' frames, staggering loads rising to twice their own height. Mothers carried puzzled, doe-eyed children and avoided the soldiers' glances as they stood back.

As we bumped our way north the valleys became narrower, their sides steep, then precipitous, until we were driving through gorges where streams tumbled and splashed across the crude rutted tracks. There was less dust, but it was hot and steamy, nearer to jungle conditions. The British Brigade was in the front line just south of the Imjin River, and what was left of the forward battalions had regrouped. Companies were understrength, as we were among the first of the reinforcements flown out rapidly from Europe. Both the Glosters and the Ulster Rifles were back almost in the same positions they held at the end of April when the main force of the Chinese spring offensive hit them in a drive towards Seoul. Since then, counter-attacks along the whole front, assisted by heavy air strikes, had forced the Chinese and North Koreans to withdraw north of the Imjin River. The two sides had almost broken contact and the war had moved to a phase of patrolling the no-man's land on either side of the Imjin

River. But the tactics of any future battles were clear.

Near the 38th Parallel, Korea's mountains make a lunar land-scape. The hills and peaks stretch away on all sides to the distant horizon. Only the valley of the Imjin River, with its tributaries snaking north and south from narrow gorges, with here and there miniature plains and paddy fields, relieve the scene. The main tactical lessons of the first phase of the war had been learned the hard way. Among the forward troops, the war was now a battle for the ridges and the peaks. Those who held the high ground dominated. Those who were road-bound or who assumed that advancing up the narrow valleys would win the war would be making—as the American and British commanders had already made—grave tactical errors. Reserve units and some headquarters could remain in the valleys and gorges if they were well protected from the front, but forward infantry had to hold the crests.

The truck dropped us at the tactical headquarters of the Ulster Rifles, hidden beneath trees across a stream in a narrow gorge. Two or three camouflaged tents and a trailer with a tall wireless mast were grouped about a small clearing. Some jeeps and a truck stood beneath camouflage nets. Inside the main tent, its sides drawn up to let what breeze there was circulate, we met the CO, 2IC, adjutant and two or three other officers from head quarters. It was just before lunch and paraffin burners roared somewhere outside. There was some Japanese beer in a hole in the ground, kept cool by a trap-door, and it tasted good. Intro-ductions were carried out by the adjutant, with studied courtesy all round. The newcomers were offered two of the battered seats borrowed from some Seoul hotel or municipal hall. We declined and sat on the ammunition boxes sprinkled about.

The officers looked weary, but not dispirited. Their words were measured, economical, matter-of-fact. There was little humour, no sentiment. During the next few weeks I heard about the April battle, and how the British Brigade had been quickly surrounded because the South Korean Division on the flank had 'bugged out' almost immediately. From battalions each number-ing about 600 officers and men, the Glosters (holding the most

L

vulnerable position) emerged with 5 officers and 34 Other Ranks, the RUR with about 14 officers and 200 men. The Northumberland Fusiliers, partly in reserve at the time, had fared better.

The main Chinese tactic was the 'human wave' technique of assault. Their vast numbers allowed this, and the Chinese force in the April battle was estimated at 60,000. If the front line was gunned down, the next 'wave' simply stepped forward in its place, and so on... Their infantry could march twenty miles a day on only a bag of rice. They attacked at night, filtered round to the rear of their enemy's position, then made an unearthly din with bugles and klaxons to create panic among the enemy's soldiers. There was nothing in British tactics manuals on these battle techniques.

After a lunch of corned beef stew, the adjutant told us our allotted companies. I was to take over an officerless platoon in D Company. The jeep took me forward, and I found D Company cleaning their weapons under a crude shelter made of branches and leaves at the base of a hill. They were all bronzed, stripped to the waist, their jungle green bleached to a pale olive. The officers carried gnarled walking sticks, cut from branches and polished with boot polish, like an Irish blackthorn. D Company commander was Major Gaffikin, or 'Gaffie' to the company, a slender man with a Baden-Powell thumb stick and a sad face like a disappointed basset hound's. But he was a warm-hearted, intelligent man, with a quick lurking humour ready to despatch the shadow stamped on his face by the bloody battles of January and April. I heard later that he had been awarded the DSO for leading what was left of his company through a gorge as the Chinese stood above, picking them off one by one. Gaffie was the last one out of the gorge.

The other subalterns in the company were friendly and welcoming. Two of them had got MCs from the earlier battles. As the afternoon gave out and the sun went down every man in the company began to make his way to the company position at the top of a mountain. We had the highest position along this section of the front and it was a long climb, about 2,500 feet, but with magnificent views from the top. To the north, a loop of the

Imjin River lay like a silver snake, coiled in the evening sun. Beyond the river, jagged mountains vanished into North Korea.

Our positions were on the crests of the highest hills just south of the Imjin. The men shared small American 'pup' tents— simple, dark green canvas affairs. Junior officers had one each, and company commanders a slightly more luxurious British tent. Some of the men had constructed dug-outs beneath their tents. There were forward trenches just below our positions and guards were posted to the front, rear and flanks of the main firing trenches. The subalterns took a turn on guard, which worked out at four hours on, four hours off each night. Reveille was soon after first light. You soon adjusted to four hours' sleep, sometimes less. The morning descent to the plain below took about an hour, and the men came down individually after taking their breakfast of 'C' rations in their dug-outs.

These were American rations, supplied in 24-hour packs. The cardboard box was stamped 'Ration, Individual, Combat, C-3, Menu No—'. The particular menu contained was a matter of chance, not choice. But in any case there was little variety, and after a few months you were intensely bored with the daily problem of attempting to vary the menu. The several small tins contained soup, beans, frankfurters, and preserved fruits—perhaps cherries, peaches, or sliced pineapple. One tin contained packets of powdered coffee, sugar, and two round cakes of cocoa, which dissolved after a good deal of prodding. Another contained a miniature tin stand, wax discs and matches, for heating up the main course.

After a month or so most of us had some mild form of dysentery, probably from the local water in the streams. Bowel disorders followed as the digestive organs revolted against the sameness of the food. A good deal of swapping went on, to cover individual tastes or allergies. The quartermaster managed to ferry up supplies of tea for our brew-up twice a day, but this was the only concession to English tastes.

The immediate and most obvious difference between peacetime soldiering and front line conditions was that one shared the life of the men almost totally. The apartheid of officers' mess and other ranks' billets and canteens no longer existed. Nominally, one was entitled to a batman, but there was almost nothing he could do: no equipment to clean, no beds to make or rooms to keep tidy. The officers oiled and stripped their own pistols daily, and kept their own boots preserved (as distinct from shined) with boot polish. Again, with the shortage of riflemen, one could hardly add to the duties of any one soldier by requiring him to look after the platoon commander's minor needs.

I had a dug-out to myself, however, where most of the men shared one, and when the green poncho was tugged over the top it was possible to switch on an electric torch to read. During the night, field mice emerged inside the dug-out, sometimes perching on one's knees in the darkness or, more disturbingly, darting across the face.

There was little activity along the front at this period, and even shelling had died down. Dawn was often exquisitely beautiful. It was easy to see why Korea's oriental name is 'land of the morning calm'. As the sun came up, the valleys below were blotted out by low-lying white clouds, blanketing the paddy fields, the crests and peaks of the hills rising from a cotton-wool sea. Everything was silent. Then the clouds would lift and dissolve, and the cicadas began their ceaseless drumming in the trees as the heat of the day came on. Now and then, in the valleys, you could see not one magpie, not two, but a dozen or so, fluttering among the fir trees with their long dipping tails. In other circumstances, at other times, the sheer beauty of it all, like a delicate silk screen, would have been consuming; but in the valleys the machinery of war interrupted the eye.

An all-day probing patrol across the Imjin River was announced for D Company two days after I arrived. The battalion was to take two companies across, form a base, and then send one company ahead, thinning down to a platoon, then to a section, and finally half a section of three or four men to take a look just behind a pointed hill feature some miles north of the Imjin, where field observers said there were enemy mortars. D Company would be forward. I discovered that my platoon would form the apex of the probe and that I was selected to complete the probe, with one corporal and two men.

There was nothing unfair in this, but momentarily I was surprised that the tactical plan should call on someone with so little experience in the field. These deep probes were very rare at this time, and when, a month later, my Border Regiment colleague Ken Hodgkins was killed on a very similar probe, I wondered at the tactical wisdom of putting inexperienced subalterns, new to Korean conditions, to such a specialised task. Yet looked at fairly, it was right that one of the seconded newcomers should be selected for this patrol. The Ulster Rifles had lost half its subaltern strength in two major battles.

For the first time I experienced the real responsibilities of minor command as I studied the maps and decided on a tactical plan for getting behind the conical hill north of the Imjin. We had good maps, showing hill features, valleys, gradients, the villages, and main streams flowing north and south into the Imjin River. The maps were American-supplied and said much for the thoroughness of their aerial reconnaissance.

The two companies crossed the Imjin at the shallow reach already christened 'Gloster crossing' from the April battles, and made a beach-head in the low ground on the north banks of the river. From here, the long snake of the main patrol went north, guided by two or three Korean scouts attached to every battalion. These were remarkable men, usually in their early twenties, educated at Seoul, some of them graduates of the university there. The main advantage was that they spoke English and could interpret when we passed through a village and asked the local people about the activities of the North Korean forces. They were also cheerful, helpful, self-reliant, and were excellent for morale among our own troops, as they had the Koreans' extraordinary ability to carry immensely heavy loads on the march.

The brigade also had on call a small army of Korean porters, always reeking of garlic, few of whom understood English, but who were nevertheless willing workers, repairing the dirt roads or manhandling a trapped lorry out of the mud after a rain storm.

Our route lay across paddy fields as the ground rose, and, as
narrow mud walls crumbled, we marched direct, up to the
knees in water, occasionally up to the waist when the main irri-
gation canals barred our path. Once on the high ground, the
squelching noises to front and rear ceased. We gradually thinned
down to one platoon as we moved forward. I could see the hill
feature about a mile to the north now, its sides pockmarked
by artillery and mortar shells. It looked desolate and deserted,
but intelligence reports said that the enemy were below the crest
on the other side.

It was still ominously silent as we came within a few hundred
yards of the cone. We were now reduced to a section of ten men
and I wondered if I was being lured into an ambush or a trap.
There was no way of knowing, as the Chinese and North Koreans
were masters of the art of breaking contact and vanishing suddenly, as if the ground had swallowed them up. Now and then in
the yellow earth we saw telltale footprints, slightly unearthly, as
the Chinese wore a soft-soled boot with the big toe separated from
the rest, giving the footprint a cloven appearance, an animal
rather than a human form.

I was preparing the last stage of our reconnaissance round the
cone with the map and compass when a message came through on
the wireless set. The signaller gave the code name of the adjutant,
and I heard his voice, distant and unnaturally British in these
strange surroundings. The final stage of the patrol had been
cancelled. Too dicey. I was to make what observations I could
from present position, then rejoin the main force. They had me
in their binoculars, and I was covered for withdrawal by machine
gun and mortar.

So my first patrol was an anti-climax, but this was just as well:
air reconnaissance brought news via brigade headquarters that
the enemy on the other side of the cone were in strength, like
ants, digging defensive positions.

We continued patrolling across the Imjin, the companies taking
their turn, but there was a lull in the war, both armies re-group-
ing, mending their supply lines, and breaking contact except for

these patrols in no-man's land. The rumour was that early autumn would bring new offensives, perhaps soon after the summer rains had dried out. These were expected in August, and veterans said they were heavy, more like monsoons. With air superiority, the UN forces had the better tactical position during this period. The only sign of air activity from the north came at night, when a lone plane cruised overhead, broadcasting messages in English, obviously aimed at lowering our morale. Most of it was completely inaudible, garbled by the warm eddies of air rising from the valleys and hills. Now and then the lone flier dropped some sort of light bomb, not much bigger than a hand-grenade, before turning away. The troops became quite attached to these visits, more like a courtesy call, and an odd sort of affection grew up for the familiar noise breaking the total silence of the night. The plane was christened 'bed check Charlie' by the Americans, and if he missed us for a few nights we almost resented the departure from a familiar routine.

Time began to hang on our hands when we were not patrolling, and to escape from the sun during the day we constructed shelters, called 'bashers' for some reason, using branches and sometimes the timbers of deserted Korean houses. These became quite elaborate affairs, and national servicemen from the building trade brought out their skills, climbing about the roofs like monkeys, lashing timbers and suspending bamboo walls to give a cool shade inside. Latrines also became civilised places, the throne supported on two joists and made from an upturned wooden packing case with two central slats taken out, with soil packed around the base, hiding the cavern below. These ambitious constructions were enclosed within a cool cabin of branches and leaves, usually on a slight knoll with a fine view of a valley. Hanging your jungle hat on the door served as an 'engaged' sign. It was an excellent spot for deep thought.

After a month or so, I had to take over another platoon in the adjacent company. Their platoon commander was leaving, his service completed. The platoon was in an exposed position, covering a small hill in the middle of the twisting gorge just where it made a final bend to the Imjin valley. A few days later, when checking some grid bearings on the contoured maps, I discovered that my platoon's location was immediately beneath and just forward of Hill 235, where the Glosters had made their final stand in the April battle. The discovery brought some mixed reactions which I kept to myself.

The company commander had to make a detour to get across from the steep escarpment where his other platoons were positioned, as this was the extreme left flank of the battalion. Perhaps remoteness had affected morale in the platoon. There was a good deal of litter scattered about the men's dug-outs—just the sort of bright bric-a-brac to attract artillery spotters or air spotters. Guards were posted at the base as well as the crest of the hill, as this was almost certainly a route for infiltration to the south at night. When I visited the guard at the base of the hill during the night, they were not alert. At the top of the hill it was obviously routine to fall asleep until the next guard disturbed you.

Next day, checking over the positions and equipment, I discovered that one of the two young Korean porters attached to the platoon was sharing a tent with the cook, a grey-faced, shifty man who made himself scarce throughout the day between brewing up tea and producing tasteless corned beef stew from some British rations now coming up supply lines. The other young Korean—both of them in their teens—was sharing a tent with the oldest soldier in the platoon. The arrangement seemed to me an odd one, as there were two spare tents with the platoon, and the Koreans were perfectly friendly with each other.

It was a delicate situation and I thought it best to discover what the other men thought of it. I got the answer from one of the regular soldiers, an Ulsterman who had been through the whole campaign and had seen more action than most of the men there. I asked him what was his attitude to the Koreans sharing a tent with the cook and an old soldier.

'Sir,' he said, 'any man that's been here above six months, never setting eyes on a woman, begins to fancy these young boys. I've fancied them myself, odd moments. But either we all bugger them or nobody does.'

I moved the two Koreans into a tent together. They seemed
cheerful enough about the new arrangements. One of them
seemed relieved, so far as it was possible to tell, but oriental in-
scrutability did not make for certainty.

It was not difficult to discover why the morale of the platoon
had sunk so low. They sat around all day doing nothing. Casting
about, I managed to get a football from the quartermaster, and
we constructed a volley ball court just off the road below. A long
piece of camouflage netting came in handy, and pounding boots
soon made the court perfectly good. We also constructed a miniature swimming pool by damming the stream with sandbags. An
initial hazard was the discovery of two dead Chinese and a mule,
obviously from the April battle, decomposing in the undergrowth.
This explained the disagreeable smell hanging about the river at
this point, always pungent on the night air. We were glad to bury
the smell along with the corpses.

Once or twice I paid a visit to the Glosters company on our left
flank, covering almost the same position they had occupied in
April. They were hospitable, and on one or two evenings as the
sun went down I took a nip of brandy with them. They were on
a ridge above and behind my position, looking down at the round
knoll where my platoon were perched, covering a sharp bend in
the narrow valley. It looked very exposed.

One day I had a surprise visit from the Brigade Commander,
Tom Brodie. He arrived with the battalion CO, the 2IC, adjutant
and my company commander. Brodie was a friendly, direct man,
with an extraordinary memory for faces. The next time I met him
was in Ismailia at Suez, about ten months later, where he noticed
the Korean ribbons on my khaki shirt and said crisply, 'I remember you. Left-hand platoon, Baker Company, Ulster Rifles, Solma-
ri, Imjin River. Right?' He was right.

The purpose of this sudden visit to my platoon, I discovered,
was to re-position the battalion, giving it greater depth by concentrating on the higher ground. My hill was to be abandoned
and heavily mined, as it was isolated and much lower than its
neighbours, with little strategic value. In fact a muttered aside
between two senior officers seemed to confirm that the position

was really suicidal, so far as regular Chinese tactics were concerned. I did not tell the men, but in any case they were not sorry to leave. Their common sense had probably told them that the nimble Chinese and North Koreans could have sprinted over our pimple in a matter of minutes. We left plenty of barbed wire to delay or frustrate any further tactical use of the hill, and the engineers hung their red triangles about the perimeter wire once they had hidden their mines.

The companies on the neighbouring hill changed personnel as another group of national servicemen went off and an incoming group joined us. Each platoon received its quota of new arrivals. The newcomers were always instantly recognisable by their pale skins and green uniforms, unbleached by sun, untorn and unpatched. They asked lots of questions, slightly nervously, some of them picking hopefully at rumours from Pusan that the Ulster Rifles and the Glosters were to be withdrawn soon.

The Twenty-Ninth Brigade was being expanded by other British, Australian and Canadian battalions to form the Commonwealth Division under Major-General Cassells. Rumours of an autumn offensive persisted, but static armies in the field always foster rumours. As July gave out, we began to patrol in greater strength. The hot, dry summer had reduced the Imjin to a width rather less than the Thames at Westminster, and shallow enough to let tracked vehicles pick their way across at the broad curve near Gloster Crossing. With this additional support, including medium machine guns and mortars, our patrols reached further into no-man's land, always pulling back to our own lines at night. One day in August, however, when we were across the Imjin in two company strength, we were caught by floods. All morning the sky was overcast, and at mid-afternoon the rain came down. The monsoons were expected soon, but what happened caught everyone by surprise. No one had foreseen this downpour, which went on and on, increasing in its intensity, until it roared like a cataract about our ears. There was no sheltering from it: the rain sheeted down through the trees, and the dusty track beneath our feet became a stream.

All patrols were called in quickly and we prepared to return across what had been a waist-deep ford. When we got back to the Imjin there was an unpleasant shock. It had risen about ten feet and was now a gushing torrent. The far bank was invisible in the downpour. Wireless contact was imposible because of the roar and the interference of the electric storm. There was no question of getting across that day. We took up defensive positions for the night, though even these seemed ludicrous as the rain continued through the whole night. I posted men at various positions, instructing them to cover an angle of fire—which meant pointing one's arm out into the black downpour and shouting the angle to be covered. We lay on our bellies, most of us in three or four inches of water. For a time I curled up in the water and slept.

The rain stopped near dawn, but we began to wonder if we would ever get back. A report which J. L. Hays wrote in the *Sunday Times* soon after remarked:

> The fast-running Imjin broadened from 150 yards to 300 yards. One ford which had been thigh deep at sunset, was 19 feet in depth at dawn. . . A prolonged stay over the river among the enemy was not expected. Our troops carried little food and ammunition, or bad weather clothing or equipment. But then, as they described it, 'the roof fell in.'

We now learned that other patrols were caught across the river, including the Belgians on our flank. The greatest mystery so far was why the enemy had not attacked an isolated, almost defenceless force. We had not eaten for twenty hours or so, and wondered when an air drop might come. The torrent behind us made any river crossing that day out of the question. In mid-afternoon, American transport planes appeared overhead and cartons fluttered down, bursting as they hit the ground. For some extraordinary reason the first air drop contained scores of boxes of boots. The 2IC of the battalion got on to the brigade wireless net and said some crisp words. Later in the afternoon, another air drop arrived. This time the cartons contained 'C' rations, and the men brewed up their tins of soup, beans and frankfurters and then coffee. Everyone felt better. More ammunition was dropped as the light faded.

The 2IC gathered us together and passed what information he had. The pace of the river made a pontoon bridge out of the question: it would be swept away like matchwood. The engineers hoped to get a cable across in the next twenty-four hours. We would spend another night on the north bank, moving out at last light to occupy the high ground, about half a mile to the north. Night patrols would go out; mines would be laid around defensive positions; movements must be strictly controlled.

The men were obviously depressed by the news, and there were grumbles about the engineers not being around when they were needed. But one look at the Imjin—with uprooted trees, the remains of peasants' houses, and now and then dead cattle, turning over and over in the swell—showed that these were freak conditions. And one already knew this characteristic of the British soldier: a preliminary grumble, a few thick curses, then he put it out of his mind and carried on with the job in hand, cheerfulness breaking through.

By now I knew I was unusually lucky in the make-up of my platoon. Ulstermen formed the main part of it and the grubby Field Message Book surviving from those days brings back some of the names, and the weapons they carried: Murphy, O'Kane, McCulloch (2in mortar), Casey, Pierce, Deagan and Spear (bazooka), Birch (wireless), Roddy, Varley, Hooke, Sinfield. The twenty-eight men included two or three veterans—quiet men in their forties with soft Ulster accents, who had come through all the battles and were a calming influence on the rest. Among the new national servicemen I had two cockneys, who were either barrow boys at home, or gave a perfect imitation of being so. They were our resident comics, knew their function, and played the part, a sort of music-hall act which kept the platoon's spirits high. They even wore their jungle green hats nipped into rakish trilbies, with a Sid Field effect. When they settled down to their 'C' rations, sitting close together, they kept up a scornful and inventive patter.

'Cor, what you got on your menu tonight Joe? See 'ere, salmon, caviar, steak, pickled onion, tomato sauce, bread 'n butter, loverly.' He held aloft the familiar frankfurter speared on his fork.

Morale was no longer a problem with men like this in the platoon. It was curious to think that at school they might have been the despair of a headmaster. We had to draw on this reserve of high spirits over the river, and particularly on the second night. When the 2IC announced the details of patrols, I had to take my whole platoon forward as a mobile battle patrol, not so much to provoke the enemy as to give early warning if there was any concerted movement towards our main position. If I made contact with any enemy patrols, I would try to seize two prisoners and bring them back for interrogation. Divisional headquarters were still uncertain of the strength and identity of enemy units to the north.

The men groaned, had their grumble, and then set to, cleaning and oiling their weapons, squeezing their wet socks for the umpteenth time. We set off towards midnight. The sky had cleared now, and a pale moon hung above. Night patrols were never pleasant as the gradients were confusing at night, even with a compass. It was easy to get lost among the convolutions of the ground and head towards the enemy lines even when you felt certain you were heading for your own. We kept to the high ground, crouching across the bare ridges like thieves stealing away from a crime.

After about two hours we were fired on, but not from near at hand: a machine gun chattered across a valley. The bullets thudded in a scattered pattern. We went to ground. It could be a forward observation post, a solitary sniper in a dug-out, a patrol like ourselves, or the main enemy force. There was no further fire. The silence was complete. After ten minutes or so, from a flanking dummy position, I lobbed a flare from our platoon mortar. The landscape was bathed in pink light, sending shadows darting across the rocks and shrubs before it gave out. Silence again. I reported back through the company wireless link. The 2IC ordered me to break contact. If the enemy were in large numbers, they might come south in pursuit and catch the rest of us across the river.

We withdrew, keeping near the ridge where the moon threw shadows. There was a strange unearthly beauty to the landscape, silent and undulating, the light clear and the air warm, with a slight moistness clinging from the damp earth. We avoided the villages and hamlets nestling in the clefts of the ridged hills, with the exception of one, where it was part of our mission to listen for signs of activity. Various intelligence reports had suggested that this particular hamlet was visited by North Korean soldiers, moving in by night to observe our positions from cover during the day, then withdrawing by night again a few days later.

We lay down above the village for an hour or so, listening for any signs of activity. There was none, but our orders were to move through the village in any case, to show our presence to the villagers. I divided the platoon into two sections, and we moved quietly on either side of the little hamlet. I was leading one of the sections and coming across a small compound towards one particular cottage when I had the most frightening experience during my time in Korea. Moonlight and some shadows played a trick with a low bush directly in my path. I was certain that it was an armed man, crouching and taking careful aim. Ridiculously, I did not wish to lose face with the corporal and the other men behind me by going to ground in front of what was possibly, after all, an innocent bush with a stray branch pointing towards me. Besides, it was too close to duck now: he was aiming between my eyes. I felt the hair rising sharply on my neck and went cold, certain that—in the army's phrase—my 'number was up'. With four or five yards to go, the projecting branch and the low bush came into sharper relief, showing their true form. It was one of those moments, common enough in war, where farce and disaster were closely juxtaposed.

We got back to our main positions by first light and during that day the engineers managed to get a cable across the Imjin, firing a projectile to straddle the broad span of the river. The tactics of ferrying troops by helicopter had not yet been developed and the few light helicopters available were reserved for urgent casualty cases. The river had gone down, however, and by dusk that day we were all across on the precarious cable ferry the engineers had rigged up. For some time most of us had twinges of rheumatism in the legs or the hips, but our dug-outs and 'pup' tents seemed astonishingly luxurious after our soaking across the river.

As we dried out, the battalion was given a short period of rest from active patrolling.

Now and then there were visitors—usually journalists, carrying battered typewriters festooned with luggage tallies. Eric Linklater called by, asked intelligent and perceptive questions, and later produced a book. As the officers talked, the grim experiences of January and April unfolded. In January, when an arctic wind swept down from Mongolia, the battalion had been in an exposed position on the Han River. The ordinary winter equipment issued to the British troops had been tragically inadequate during the first winter. The ice-laden wind cut through greatcoats, shirts and string vests as a stiff breeze passes through chicken wire. Frostbite on hands and feet was often severe, and if the men did not keep constantly on the move—even when lying down in defensive positions—they lost fingers and toes. One of the most trying things for an officer in these conditions had been to hear men whimpering with pain like infants, but with no means of putting things right. The equipment branch of the War Office had blundered, in fact. By the second winter, thick parka suits and double-thickness gloves replaced the pathetic jumpers and woollen mittens of the first winter.

CHAPTER 18

Autumn Offensive

Irish regiments seemed to attract unusual characters. Even in the line the battalion had its share of eccentrics. One subaltern, Houston Shaw-Stewart, endowed with a rich, doting mother, received stoutly packed parcels from Fortnum & Mason's about once a month. He was very generous with these, and passed around jars of chicken in aspic, gentleman's relish or crystallised ginger. 'I say chaps, look here,' he would shout, as though sharing round a tuck hamper in a dormitory at school. He was more fond of gin than most of his colleagues, but had reserves of courage and had won an MC by leading his men down a defile, swinging impatiently at the bullets with his walking stick as he strolled along, like a squire visiting his estate at twilight.

The padre attached to the battalion, Ronnie Woods, was a tall, athletic man with long sideburns linked to tufts on his cheekbones. He walked the various hills with astonishing speed, constantly visiting the different companies. Somehow he managed to convince us that we were all very lucky to have such a healthy, out-of-door life, with clean air and none of your city smoke. The medical officer, John Craig, was an extremely handsome, fairly young man, probably irresistible to women and given to deploring their absence in the line.

The quartermaster had his stores well south of our positions, just north of Seoul. I paid a visit one day to stock up with equipment for my platoon and found him installed in a bombed school, with big cartons of woollen socks spread around the compound. He waved his arms generously towards the cartons and I travelled around, dipping into these capacious bran tubs, dumping piles of stuff into the jeep. The QM had requisitioned a neat ménage out of one end of the building, and two young serving girls brought tea. His slippers were tucked under the bed on a carpet. The big high-powered wireless set by his bed looked American and new, probably part of the exchange of war.

There was a good deal of barter in Korea as the Americans poured in their supplies and matériel with prodigal abundance. Within a few weeks I had acquired—as most junior officers did— an American M1 carbine, a much more useful weapon than the officer's pistol with which we were supposed to attack enemy bunkers. The American carbine could fire automatically, had a slightly smaller bore than the British .303 rifle, and above all was lighter, handier, and almost as accurate. Ammunition was easy to come by, and there seemed to be plenty of it with any battalion in the line. An unofficial market pricing system operated, and since the British were allowed a small quantity of spirits in the front line through NAAFI channels, whereas liquor was officially forbidden to American troops at 'the sharp end', a type of black market operated. A carbine could be exchanged for a bottle of gin. A case of whisky might be exchanged for a battered jeep from a base camp, or for a new engine from American MT stores. The dislocations and wastages of war, coupled with the endless bounty of Uncle Sam, made these transactions unremarkable. The real black markets operated in Pusan and Seoul, where street stalls and shops sold liquor and cigarettes, American army wristwatches, binoculars, cameras, boots and shoes, towels, and a variety of pistols, ranging from German lugers to the old fashioned six shooter, which some Americans favoured as personal armament when travelling about.

Inevitably, Korean houses were often stripped of ornaments and useful articles of furniture as armies moved through. Brass pots, bowls, even delicate carved screens would be slung on to military vehicles if they could contribute some minor comforts further up the line. Here and there, in a battalion headquarters mess or a senior officer's dug-out, elegant carved screens rose incongruously from sandbagged revetments.

September came, and the war died down to a whisper. Even patrols failed to make contact with the enemy. I heard from the

M

adjutant that my turn had come up for 'R and R' leave. This meant five days in Tokyo, on a quick flight direct from the American air force base at Seoul. The 'R and R' was supposedly 'Rest and Recuperation', but the Ulster Rifles officers who had already had their leave hinted that after five days in the fleshpots and brothels of Tokyo one came back to the line in order to recoup one's strength. I wanted to see Tokyo and, if possible, aspects of the older culture of Japan. The war machine at Kure and Hara Mura had given me no glimpses of this. Also I wanted some solitude, or at least the anonymity of a big city again, away from people I knew, removed from military routine.

Catching an American plane was a haphazard affair. Most of the planes taking off from Seoul were fighters, with napalm canisters fastened like plump silver sharks beneath the wings, heading north. The air traffic office was in a hectic state. A lone British officer wanting to hop on to the next transport plane to Tokyo, even with the required military pass from divisional headquarters, was merely an additional loose thread. My call to the plane was sudden and succinct, from a man chewing a cigar.

'Where's that British officer? Okay, there's your plane. Out there. Hop on. He's ready to take off.'

I sprinted across the tarmac where fighter jets blasted the ears and whipped at the fresh jungle green outfit I had picked up at the quartermaster's stores on the way to Seoul. There were only two other men in the freight plane, both Americans, and we sat silently on benches against the fuselage, engrossed in our own thoughts. It was cloudy over Japan and the pilot lost his way. We flew around for an extra two hours, dipping through holes in the clouds and then banking up again sharply as the navigator scratched his head. The cockpit door was open, and so far as I could tell he was using an ordinary small-scale map. I wondered how much fuel we had left. Finally we nosed down at Tachikawa, and an hour or so later I was having the extraordinary luxury of a hot shower, standing on cool tiles between white tiled walls, with clean towels over the shining rails.

It was an officers' hotel, and in retrospect one concludes that the smiling girls who came to draw back the sheets, to collect

shoes, to bring them back, or to clean the wash basin, or change
the towels suddenly, or dust the telephone, were available and on
call. Odd that the surmise did not occur at the time. Perhaps it
was the old, nagging fear of catching 'a dose', as everyone put it.
I went out and walked down the Ginza, the main street, looking
into the shops, seeing the destruction of a city by neon, ham-
burger stands, strip joints, and the towering insult of 'The Ernie
Pyle Theatre', swaddled with neon and garish posters. It was
just as they said: Japan had joined the West and Tokyo was
already outdoing America in the neon culture pasted on to the
main shopping streets. A Japanese boy tugged at my sleeve as I
walked down the Ginza:

'Say, mister, you wanna meet my sister? She's sixteen. . .'

I had no guide to tell me how to get out into the country dis-
tricts where some of the older Japanese culture might still be
preserved. Language was an insuperable barrier immediately one
stepped outside the hotel or the more obvious restaurants which
had long gone over to a mongrel mixture of American-Japanese
food. The tropical green uniform marked me as an alien, one of
the many confusing but equally corrupting military occupation
forces, who spawned pimps, prostitutes and touts in the Ginza.
You could not blame the glassy, almost sightless look which so
many of the middle-aged Japanese gave as you threaded your way
down a side street, looking for signs of an older Japan. Now and
then I caught sight of people bowing and bobbing to each other
as they took their leave at the end of a casual meeting, but the
sight of a Western stranger in uniform cut short the courtesies
and they would hurry off. I was glad when the five days ended
and I returned to Seoul. Only the piece of black chiffon, which
became a scarf, provided a delicate memory to show that lovers
can cut through cultures.

The stalemate in the war continued. Truce talks began at Kaesong. Initially the soldiers in the front line had high hopes that
the war would end in a matter of weeks or even days. After all,
if the two sides had agreed to meet in order to discuss a truce, all
that was needed was a simple ceasefire agreement, to be followed

by the repatriation of prisoners of war. But these simple assumptions did not hold in the ideological clash centred on the dusty compound at Kaesong and later Panmunjom, as the delegations took it in turns to walk out. Word soon filtered through to us that the talks were bogged down in wrangles about procedure, then about who was to blame for the war, then about the terms for repatriation—voluntary or forced—of prisoners on both sides.

As autumn approached, our soldiers began to dismiss the peace talks with blunt epithets. Then word came through that we were to advance across the Imjin to form new positions. The whole of the Commonweath Division would move north of the river about four or five miles. The news puzzled the ordinary infantrymen, who did not think this the best way to improve chances of a truce. Later, one discovered in accounts of the diplomacy and strategy of the war that the UN authorities were seeking better bargaining positions for the talks, since the chief stumbling block in the negotiations seemed to be the correct line for a ceasefire. In the western sector, Line Kansas lay south of the Imjin River and the Thirty-Eighth Parallel. The new positions would be a few miles north of the parallel.

The political strategy did not filter through to the men in the dug-outs on Line Kansas, and if it had they would not have been impressed. There were one or two grumbles as orders for the advance north came through from divisional and then battalion headquarters. By now the men had made themselves comparatively comfortable, and fresh food was coming up on the supply lines. The summer stalemate had allowed roads to be improved. If the advance north across the Imjin meant going back to 'C' rations, the men were against it. It was as simple as that. For the veterans, those who had come through the bitter January and April battles, it was also a nerve-racking development. They had seen so many of their friends killed that they counted themselves lucky so far. Moreover, with rumours already circulating that the Glosters, the Ulster Rifles and the Northumberland Fusiliers were soon to be withdrawn from Korea, these new plans seemed like a betrayal. In personal terms, infantrymen wondered whether this time their 'number would come up'. One or two of the older

men in my platoon voiced their thoughts, in the familiar, jocular, self-deprecating asides, as they oiled their rifles and cleaned
magazines. On the hills, the trees and scrub showed the first tints
of autumn.

The Commonwealth Division was concentrated in the area of
Majon-Ni, where the Imjin is joined by the Sami-Chon River
from the north whilst the main stream snakes into a huge bend,
almost doubling back upon itself. We were to attack the area
lying between the Sami-Chon valley and the upper portion of the
Imjin's 'S' bend, to form a new line called Jamestown.

We left our positions on Line Kansas well before first light.
Coming down the steep hills in total darkness was hazardous,
and one of the men cut open his leg when he slipped with a
machine gun which formed one of the semi-official weapons in
my platoon's armoury. But he insisted on continuing with the
rest of us. A field dressing staunched the wound. By first light
we were at the Imjin at Gloster Crossing. There was no sign of
activity, and as we crossed, the artillery to our rear began the first
of the shelling which continued all day. By mid-morning we had
taken up temporary positions on the Yook-kol feature, familiar to
us from our patrols, and part of the no-man's land of the summer
stalemate. Our real objective lay some miles north-west on the
hills above the Sami-Chon valley.

We came under fire around the middle of the afternoon and
went to ground whilst artillery spotters called down medium
artillery and heavy mortar. The North Koreans and Chinese
were extremely accurate with their mortar fire, besides being
highly mobile, and we never felt comfortable in static positions
when they had drawn a bead. As the afternoon wore on, however, the next phase of the attack was ordered over the wireless
set, with the pre-arranged code names. A and B Company of the
Ulster Rifles attacked a feature called Taraktae. My own platoon
was forward on the left flank, and when I gave orders to the three
section commanders, I had a fleeting sense of excitement. Battle
at last. As we moved forward up a slope and the crack and thud
of fire came from the front, the feeling became more pure and,

for the first time, a terrible allure invaded the senses. Later, I recalled with an awful clarity its components. In battle, life reduces itself to a primal simplicity. None of the nagging round of worries of modern living: the state of one's bank account, bills to pay, or the emotional debts of relationships, uncertain friends, or remembered quarrels. All these dropped away as one went forward. Life became an exhilarating equation, pristine in its simplicity, uncluttered by morality. Kill or be killed. Nothing could be simpler, less compromising or bothersome. Even the landscape gained a sudden, terrible clarity in the afternoon sun, as though seen for the first time.

Then more immediate matters entered the mind again. I called out orders to the section on the flank as we moved up the hill. We were being fired on from the crests above us. A platoon to the rear gave covering fire with Bren and rifle, and I saw the bullets kick up the bunkers on the brow of the hill. When we finally charged with bayonets it was one of those anti-climaxes which so often come in war. The enemy had vanished beyond the hill, uncannily, as though the earth had opened up. The shallow slit trenches were empty, with only the telltale cloven footmarks to indicate that they had been here. We went to ground and I reported on the '88' wireless set. The order came back to stay where we were and not to advance further. The pause that followed lengthened as the evening sun itself sent long shadows over the landscape. Word came through by means of the code letters covering different phases of the attack that on the right flank of the division the Twenty-Eighth Brigade were held up on the high Kowang San feature. As night closed, we could hear artillery pounding in the east and the whine, then the stutter, as US jet fighters dived and gunned beyond the ridges of Kowang San.

We dug in, posted forward observation posts, and strung strands of barbed wire just below knee level in front of our positions in case of an enemy charge during the night. None came, and at dawn, when the usual blanket of cloud lifted from the valleys, moist and steaming as the sun rose, the land was quiet. Air reconnaissance reported no sign of the enemy, and during that day we moved forward another mile to take the ground just above a hamlet called Kunsang-Dong, where we began to dig in again. This was to be our section of the new Jamestown line, and apart from one night-time scare when a nervous platoon commander in a neighbouring company almost opened fire on a returning night patrol, the next few days were uneventful. We were back on 'C' rations again, now that supply lines were extended beyond the limits of the roads built by the engineers south of the Imjin, and the men had to work hard to create a new line of dug-outs and firing trenches in unyielding stony hillsides.

Some sporadic shelling arrived on our positions from heavy artillery well to the north, none of it very accurate as it was long-range stuff, but it quickened our trench-digging operations. We had learned to judge the arrival of the shells by timing the distant crump of the guns, followed soon after by a long whine and whistle as they came near, or passed over our head to land on positions to our rear. They varied so much in aim that I grew careless except when the whine sounded unusually close. One afternoon when our forward positions came under the enemy range finders, I was moving from one dug-out to the next, visiting the more nervous of the younger soldiers, when a shell landed close. The blast caught me half-crouching, and I must have reeled, my ears ringing and deafened. This was followed soon after by a sharp headache. As I lay in my dug-out holding my ears, I suddenly saw a stretcher outside. The platoon sergeant had ordered it, and, though I was temporarily stone deaf, I was not wounded in any way I could discover, so I was ready to order it away. But the sergeant indicated that the battalion MO had advised a check up in case there was any damage done, or a need for medication. It was ignominious to be carried away to the jeep waiting on the dirt road below, and the trip to battalion HQ made the roaring in my ears even worse. The MO did a number of tests, then wrote down a diagnosis. I was to go back to the Indian Field Ambulance Division near divisional headquarters to our rear. As I could not carry on a conversation and my temperature was up, there was nothing to do but comply.

I was put into a closed ambulance and we bumped our way south until the land became smoother and the unusual feel of

tyres on firm tarmac arrived. Another, more complete examination followed, carried out by a turbanned Sikh officer. He gazed
into my ears and eyes with various instruments, injected something into my arm, then ordered me south to a Canadian hospital
at Seoul. By this time, one or two labels flapped from the webbing
strap around my ankle. I began to feel like a lost parcel. By
nightfall I was in a low bed in a room somewhere near Seoul
University. An orderly in Canadian uniform brought food on a
tin plate.

Next day two or three further examinations followed, with
tests of my blood and sputum. By now my hearing was partly
restored, and it improved further as the day wore on. I talked
with the man in the next bed. He had lost one of his middle toes
in an accident with a carbine. He smiled craftily as he pointed to
the bandage, and, when I thought about it later, I wondered
about the 'accident'. In Korea as elsewhere, carefully planned
'accidents' with carbines sometimes happened. By aiming accurately so as to take the tip off a middle toe, the malingerer could
be out of battle for a month or two, perhaps for good.

I became intensely bored as the next day passed. The others in
the room played cards all day. I did not fancy joining in, and the
only books in the place were of the 'Miss Otis Regrets' school.
The Canadian doctors ordered me to rest, and I chafed for another two days. Perhaps they grew weary of me, or they thought
they could teach me a lesson, for word suddenly came through
that I would return to my unit almost immediately. I would
spend one night at the Norwegian field hospital, north of Seoul.
The Norwegians would make a final report next morning.

The Norwegian military hospital—called 'Normash' in military
shorthand—was housed in large tents by a grove of trees. Cool-
ness and cleanliness exuded from the white sheets and blue
blankets. The nurses were blonde and blue-eyed. As they tucked
in the blankets with smiles, my thoughts were Hemingway-esque.
It was a fine, sunny evening, and I was alone in the cool light of
the tent. I felt fully restored, and the ordinary vigour of our life
in Korea made me restless after three days' entombment in Seoul.
As the evening drew on I heard noises of cocktail party chatter

from a much larger tent across the compound. Peering out, I could see through the white mosquito screens that a party was well under way. It seemed a long, long time since I had been to a party, so I put on my green and made a detour round the trees before joining the party surreptitiously from the other direction. The Norwegian nurses had attracted a high proportion of American officers from other units, some medical, others in reserve combat units or the USAF. There were also Canadians, Belgians in berets, even one or two Turkish officers. The drink flowed freely. There was no excise duty in Korea, so the choice was liberal: schnapps, vodka, scotch, gin, or ice-cold Pilsner from crates on the floor. The canapés were excellent and I enjoyed the party, even when a nurse looked closely and said she was sure we had met before somewhere. I said that was always possible under conditions of war. 'Here today and gone tomorrow,' I said.

I crept back to my tent as the party began to thin out and jeeps bumped out of the compound. Later, my friend the blonde nurse came to take my temperature, before the final tuck-in for the night. Recognition came, and she tried to look shocked. I slept soundly. Next morning the doctor also gave me a hard look, and I recognised him from a knot I had joined briefly the evening before. He pronounced me fit and I caught a lift in a jeep going up to brigade headquarters. By early afternoon, I was back with my platoon in the Sami-Chon valley and took over from the sergeant. The war settled down to a dreary round of patrols once more.

We improved our defensive positions along the new line west of the Sami-Chon valley, laying mine fields, unfurling the clumsy reels of barbed wire in our thick mittens, wrapping the wire around and about the shanks of the angle irons spaced along our front like crude pairs of scissors pointing crazily from the scarred earth. We shored up our dug-outs and, as September brought a nip to the air, the veterans dug deeper, like moles. When I did my rounds in the middle of the night they invited me in. I would crowd down beside them, tugging the poncho across the hole, then sit, knees doubled up, with a candle glowing beside us as we talked, a nip of brandy from my officer's ration to bind their ribs. At such times, crouching beneath the earth on a parched hill under the orient moon, the land silent about us, life seemed almost benign. Questions of rank faded as these Ulstermen, some in their forties, drawn from little towns like Armagh, Ballymena or Ballyclare, recounted the incidents of the January battle at the Han River, or turned to pleasanter things, such as their quarters in the old Japanese Barracks at Yongdungpo, where the girls made them tea and giggled in the doorways.

Each of us got four hours of sleep in two snatches of two hours, between last-light and the stand-to at six in the morning, when the sun began to disperse the mists in the valleys beneath us. We found this ration of sleep adequate. One could drop off to sleep in a matter of seconds, aware that a touch on the shoulder would come in two hours' time as your turn came and the other guard crept off to his bunker.

Odd that one had never felt fitter in one's life. Every muscle fibre was tuned to perfect pitch. A day's march or three or four hours of digging and revetting weapon trenches left us glowing, not spent. There might be a temporary ache in the shoulders, one or two calluses on the palms from swinging the pick or thrusting the shovel into the stony hillside, but this vanished with the night. We were back on 'C' rations, which taxed our ingenuity as we tried to ring the changes on the menu in the six small tins. The barter system was restored, and those who hated the spiced, miniscule American frankfurters could swop for a tin of beans, or a tin of cherries for dessert. Beyond this, the quartermaster made sure that we never lacked a brew-up of scalding tea.

Brewing up in active service conditions is always an art, never a hit-or-miss affair. Most platoons or companies contained individuals known for their special abilities. In the absence of teapots or strainers, such men knew the special art of pouring the tea into the container as it began to simmer on the open fire: not too early, not too late. Any mistake in timing would bring the tea-leaves to the surface in a black tide. Usually it was the veterans who had mastered the art. They knew the different characteristics and performance of different teas, judging the size of the leaf, the exact amount required to make the aromatic brew. The 'char wallah' was thus a key man in a platoon, both priest and medicine man, ministering the tureen, presiding over the ceremony at first light and last light as we passed our tin mugs or mess tins to have them dipped into the milked and sweetened tea, delicately, so as not to disturb the sleeping leaves at the bottom. After a mug of hot, sweet tea, we could face the enemy, even face 'C' rations, with more courage.

CHAPTER 19

A View to a Death

October brought a welcome relief from the worst heat of the summer, though the days were still hot and full of sunshine. The peace talks dragged on and the soldiers in the line became bitter and cynical as the divisional newssheet contained more stories of stalemate, walk-outs, or long harangues by each delegation in turn. Then the war sharpened up slightly as both sides sought to gain bargaining positions, or tried to convey that their mood was belligerent, not defeatist.

We sent out fighting patrols by night and reconnaissance patrols by day. Ken Hodgkins, my Sandhurst contemporary, was killed on patrol. I gleaned the details from one of his corporals days later. He was moving forward with a section when he came under fire. They scrambled into a disused trench but light mortar came down with deadly accuracy. One mortar bomb landed right in the trench. I wrote to his parents, then wrote to the colonel of our regiment in London, General Shears, as I felt strongly that Ken Hodgkins's unusual devotion to the men he served should not be entirely unrecorded. Otherwise Hodgkins's death was unsung. The loss of a friend and contemporary somehow brought death very close for the first time. Other deaths and casualties had seemed the indifferent harvest of war. Now it was as if the cold stranger had stared in at the window, showing his face, reminding you that your turn might come.

Patrolling continued. The North Korean forces had withdrawn several miles, leaving a broad belt of low hills and valleys more or less as a no-man's land between the two armies. As we moved through the tiny hamlets clustered at the head of the rivulets leading off the main Sami-Chon valley, the peasants in their white clothes bent low in their paddy fields. Their attitude was clear:

hear no evil, see no evil, speak no evil. What their true feelings were we could not fathom but now and then as we approached, forward observers would report brown-uniformed figures darting away from the hamlets to the low scrub behind the villages. Occasionally we would meet sniper fire, but it would be desultory, poorly aimed, and almost beyond maximum range even for sniper rifles.

When we did observe uniformed figures sprinting away into the scrub from a hamlet, there was always a question of whether we should call for mortar fire on the hamlet, in case we ran into a nest of armed North Koreans. Most of us were against this use of fire power, chiefly on the simple humanitarian grounds that heavy mortar could devastate a hamlet in a matter of seconds, causing casualties to the innocent, without any tactical advantage gained, as the enemy invariably melted away over the receding skyline of the low hills to our north. But one day compassion did not have the upper hand.

It was a hot morning, a sort of Indian summer, with the sun glaring down as cruelly as it had during July and August. We were in company strength, with a battery of heavy mortar on call by radio and heavily armed for an all-day patrol deep into no-man's land. As we moved through various hamlets we searched the shelters which the peasants had now dug into the earth near their houses. This was where they took refuge during artillery or mortar exchanges along the front. Towards noon, with the sun blazing hot, we heard the crack of a sniper rifle off to our flank. We went to ground, but as usual it was not repeated. Once the lone sniper knew that our binoculars were searching the ground, awaiting the telltale puff of smoke, he invariably vanished. Probably his job was no more than to announce a token presence. We continued to patrol, uncomfortable in the heat, each individual carrying his share of extra ammunition, or Bren gun cases, plus pick, shovel or a medical pack and 'C' rations. Suddenly there was another flurry forward and the order came to go to ground again. I left my platoon with the sergeant and joined the forward observation group crouched in the scrub, with our binoculars trained on a hamlet lying in a cleft at the head of our valley.

Through the binoculars I could see a small compound and one or two children playing in the dust. In the paddy fields nearby a few white-clad figures bent low over the rice crop.

One of the officers was convinced that he had seen figures in uniform scuttling over the brow of a hill as the main strength of our patrol came up the valley. There was no reason to doubt this but the number seemed to be unusually large: about a dozen instead of one or two. Someone suggested mortar fire on the area near the village to warn off any enemy as well as to show our strength. Others were against it. Heavy mortar rarely achieved pinpoint accuracy, and there was the danger of civilian casualties. But among the observation party were two who had been right through the whole campaign, including the January and April battles, when so many of their friends had been slaughtered as they tried to break out of narrow valleys. It was understandable that those who had survived two bloody campaigns were not prepared to take unnecessary risks during the closing period of their service in this inhospitable country on the other side of the world.

The decision of one particular captain was crucial. He decided to call for mortar fire on the area immediately surrounding the village. The instructions were passed on the radio set to the reserve company and mortar battery about seven hundred yards south of our position. We would wait in our present positions until the mortar fire had finished. As I went back to my platoon position, I could see that our battalion medical officer, John Craig, was unhappy about the decision. His tanned, ridiculously handsome face was troubled, and he dug his foot into the earth in irritation as he went back to join the medical orderly in the middle of the column.

About two minutes later the first crump of the mortar sounded down the valley. One or two ranging shots fell well forward and to the flank. The radio sets crackled as forward observers corrected the range. Then about a dozen mortar bombs fell close to the village. Some hit the crests of the hill just beyond. Others spurted among the paddy fields. One or two fell in the village. A column of smoke curled up from one house. Then everything was silent.

We advanced and deployed around the village, securing the crests first with machine gun sections. My platoon was forward and instructions came on the radio to move through the village. I despatched two sections to right and left and went forward through the compound. One house smouldered, its low verandah smashed, and there were scatterings of bright red earth here and there among the scrub. The only noise was the wailing of a woman, who sat at the end of another house rocking back and forwards. A tiny infant lay across her lap. The husband sat nearby, his hands covering his face as he wept silently. John Craig was just behind, and he went across quickly to the woman and her child. I watched as he lifted the infant and thought I saw him wince at what he saw. He bent down in the shade, tugging at the satchel on his hip. He worked quickly and I moved on, searching through the houses. Inside, the peasants sat silently in small groups, looking fixedly at the floor, one or two pretending to do household tasks. In the compound again I saw John Craig come across with a gesture of despair. The woman was rocking her infant in her lap once again, its body swathed in surgical dressings, crimsoned with blood. I could see that John was in a black rage. He cursed the officer who had called for the mortar fire and put his hands on his hips as he looked across at the wailing mother, the stricken father.

'Shrapnel. Ripped the kid's belly open. I've stuck its guts back in, but it won't live. No point in sending for a helicopter. Not a hope in hell. Ten minutes at most. Bloody madness, this.' He stamped away across the dusty compound.

This seemed more than enough of a bad day's work and I signalled my sections to withdraw. But one soldier pointed out a small bunker just beyond the compound which had not been searched. He looked jittery, so I told him to keep a lookout. I went across to check the bunker before rejoining the platoon. I edged my way towards the opening from the flank, keeping the barrel of my carbine well forward. The shelter was a small, crude affair, about three feet below the surface of the ground, almost concealed by branches, with a narrow opening. But sunshine filtered directly inside. There was no noise. A curious unbidden feeling seemed to confirm that there was nothing to fear as I looked inside. There was just room for the low, crude bunk. On it, a young woman lay, her face upwards towards mine as she dangled her arm casually, almost sensuously over the edge of the bunk. Her eyes looked across and were caught luminously by the sun, which threw a warm, bronze tint over her skin. Her dress was open, and one breast lay exposed to the sunbeams drifting across her.

She seemed intensely still, peaceful. I glanced along the bunk, beyond the shaft of sunlight where the motes danced in suspension. Then I saw the mangled thigh near the wall, and beyond, the remains of a foot, twisted when she had stumbled in to let her blood ebb into the ochre dust beneath the bunk. I put my hand over her heart on her naked chest. Her eyes continued to stare. Then I realised that there was no life there. Only then did I see the reproach. She was warm, and dead.

I stepped outside to the glare of the sun. The world pivoted, hard and crystalline, the colours accentuated. Every detail of the stained scene registered: the mother with the infant's body across her lap; the father rocking back and forwards, smitten to impotence on the tip-tilted verandah; the scattered pots on the bright red earth; the fine plume of blue smoke drifting from the compound; the cicadas drumming in the shade, denying that summer was done. The flash of eternity. I stood still for a few moments. I knew now that my mind must follow the arrow and trace its path, from the hunted to the hunter, and back again. That in one vital, private respect, I had done with war.

CHAPTER 20

Pusan

Word came in October that the Ulster Rifles were to be withdrawn to Hong Kong. Our positions would be taken over by the First Battalion of the Royal Norfolk Regiment, who were already moving up somewhere in the rear. Two days later we handed over. The operation was carried out smoothly. Some of the unofficial weapons which had accrued to the forward rifle companies produced puzzles for company quartermaster sergeants when they took over: American and Japanese weapons and in one case an air-cooled German machine gun which fired more rapidly and at greater range than our Bren guns. Some weapons were logged at the end of inventories, some were not. The owners parted with them reluctantly, giving them a last-minute pat as to a faithful friend at the parting of the ways.

We moved out of the line on a Sunday. We were packed by 0800 hours, and the handing over was completed by 1300 hours. The Royal Norfolk men stood out from the Ulstermen, their skins pale, their jungle green uniforms a deep olive shade. We climbed into the American troop-carrying vehicles which had delivered the Royal Norfolk. The drivers were mostly Negroes, who leaned out of their driving cabs as they backed the huge vehicles towards our stacked kit, spinning their driving wheels casually, like dodgem cars at a fairground.

Towards evening we arrived at our staging camp, somewhere near Uijongbu. It was strange to see neat, orderly rows of tents, as big as billets, with signs indicating the various offices and locations. It was called Britannia camp, probably near divisional headquarters. There were camp beds, with mattresses and blankets. There was also plenty of NAAFI beer. That night the battalion got drunk. Officers and guards turned a deaf ear to the

N

carousing after Lights Out. In any case, the officers themselves
were in different degrees of intoxication. It was one of those rare
occasions, so far as British military discipline goes, when the rules
were suspended for a brief period.

Next day we all slept late. The permanent staff at the camp
were obviously briefed to lay on late breakfasts—a sort of
extended brunch, running on to noon. Some surfaced at mid-
morning, others appeared in the early afternoon. There were
plenty of sore heads. We rested for the remainder of the day.
Elderly women appeared from nearby villages, offering to wash
our clothes in local streams. We did not have to pay them—we
had no currency in any case. The laundry women were one of
the minor luxuries provided by the authorities. But we passed
over cans of fruit from our 'C' rations and other NAAFI stores
as token payment. No young women appeared; perhaps experience had already shown that this would be asking for trouble
with troops just out of the line. Blunt forecasts were already
circulating about first priorities in Hong Kong.

We heard we would leave for Pusan on the Wednesday. On
Tuesday, Brigadier Tom Brodie came to say farewell to the
Ulster Rifles. He made a speech to the assembled battalion—
laconic, a touch of humour here and there, but little or no sentiment. The compliments he paid were obviously heartfelt. The
men gave three cheers, Brodie touched his cap, and joined us
in the mess for lunch. The quartermaster had produced cham-
pagne from somewhere, nicely chilled, a perfect touch under the
warm canvas. We drank lots of it, though the rough greasy
cuisine of Britannia camp did not set it off. We drank more than
we ate.

Next day we boarded a train at Uijongbu and stretched out
again on the crude wooden bunks designed for the casualties
of war. The long train journey to Pusan took all night with
several halts in deserted parts of the country near Taejon and
Taegu. There was still guerrilla activity in some areas and we
lay stretched out on the planks in the blackness of the night,
hoping our profiles presented slim targets.

Pusan was still crowded with refugees. Hundreds sat about

the main railway station or camped without shelter on rusted, disused portions of the track. As our train drew to a halt, the children ran alongside, palms upwards. Their pidgin-English was American in its idiom and pronunciation.

'Gimme chow chow, Mister. Gimme cigarettes. Gimme a dime.'

The black market in Pusan was flourishing. We saw the stalls with their packs of American cigarettes, cameras, and bottles of liquor, the contents probably raw spirit or wood alcohol, which turned you blind. The streets were crowded to capacity with people, with family knots here and there sitting against a wall, surrounded by their bundles. As our open truck took us to the transit camp on the outskirts of Pusan, black dust seemed to swamp the town, swirling up behind us. If the Koreans bothered to look at us at all their faces were expressionless or resentful. At one point we were held up in a narrow street, crowded like an Arab bazaar with people. The driver honked impatiently. Then an old woman walked in front of the lorry and tugged off the meagre piece of cotton which covered her. She stood in front of the truck in her raddled nakedness, gesticulating. Someone drew her away into the shadows.

The transit camp was surrounded with barbed wire and heavily guarded. We were given beds in Nissen huts and flopped down. The dust and the bumping of the lorries had made us much more exhausted than in the line. Later we found the officers' mess, and one or two familiar faces appeared. They looked slightly embarrassed. Every army requires its base camps well removed from the lines in order to house or equip those in transit. But 'base wallah' is never a complimentary term among those back from the 'sharp end'. 'Base wallah' had connotations of comfortable beds, undisturbed sleep, hot water, showers, fresh food, and a local mistress or two.

Here in Pusan, the subalterns who had never made it to the 'sharp end' were anxious to give us a good time. Their uniforms were crisply starched, their knotted silk squares at the throat cool and chic. They produced jeeps and we sped along the road outside the transit camp to the US airbase along the bay. Our military passes at the perimeter allowed us entry. We parked on firm tarmac where jeeps were sprinkled about like the untended toys of rich children. The bar brought us back to civilisation. The choice was wide and the Korean barman stood ready to mix any cocktail we cared to name. Polished glasses, stainless steel sink and draining board, laminated bar top in cool pastel colours, leather seats, discreet lighting, a small jukebox against the wall: 'base wallahs' suffered not.

We drank cans of chilled Budweiser beer. Our host for the evening, a subaltern in the Royal Irish Fusiliers, was on friendly terms with many of the Americans who dropped in to the bar. We were introduced to one or two, and they were generous, expansive. There was no question of Ulster Rifles officers buying drinks. By mid-evening we were singing bawdy songs. One or two American nurses or WAVEs came into the bar, not at all put out by the songs, but applauding the more lewd refrains. They were women of the world, or perhaps two worlds. They worked hard, drank hard, and according to one of the Americans in our midst, 'screwed hard'. The same American told us of a cocktail party at the base, where a very British staff officer in the Guards drank a good deal but maintained his native reserve long after a hard-drinking nurse from Texas had conveyed her desire to him. The evening ended with the nurse downing her last bourbon with a flourish, slapping the officer on the back and declaring loudly:

'Well, major, I hope you got the rubbers on you. I feel like having my ass screwed off.'

With or without the rubbers, the major was propelled in the direction of the nurses' quarters.

We stayed for more than a week at the transit camp. At muster parade one morning I found that one of my lance-corporals was absent without leave. Someone said he had gone into Pusan for a drink with some friends, and vanished suddenly. Perhaps with a woman, perhaps something less pleasant. No one could tell. We never discovered what happened to him. He was not with us when the regiment finally boarded the *Empire Halliday*, a rusting tub of a troopship leaving for Hong Kong. An American military band played us away from the quayside. Sousa marches were interspersed with Glen Miller numbers. The band looked spick and span, unlike the shattered hulk of the town beyond the loading sheds. When we were a few miles out the contours of the coastline looked more delicate, and evening sun gave a warm radiance to the hills. Once more, I remembered Korea's native name: 'land of the morning calm'. From a distance it almost seemed possible.

CHAPTER 21

Fan Ling

The *Empire Halliday* was no luxury ship, but in the usual way of troopships the officers had a good deal of lounge space, insulated from the troops' quarters. During the four-day voyage we hardly saw the men. Those subalterns who acted as ship's orderly officer of the day brought back gruesome stories of the crowded troop deck. But the voyage was comparatively short and the fleshpots of Hong Kong awaited.

I sat in a corner of the lounge and sorted through the leaves of the journal I had kept. In Korea, the journal had become a series of cryptic pencilled scribblings on a Field Message Book ('Army Book 153')—a scratch pad hatched like French notepaper. It was designed for taking down orders in the field, or messages over the wireless, not for private journals. I remembered thinking as I scribbled away in my dug-out that it would probably count as a grave offence against regulations, as the information in it might be useful to the enemy. Writing had been a furtive matter, therefore, and I had to conceal the notebook in various places as the months passed. The notes had become very dogeared. Sorting them out occupied a good deal of the time on the voyage to Hong Kong.

Hong Kong island looked very green and British as we approached Kowloon harbour. The upper reaches of the Peak behind Victoria were swathed in damp mist, adding to the sense of a British presence. At the quayside we were greeted by the pipes and drums of the Argyll and Sutherland Highlanders, who had completed their Korean service some months earlier. There was a speech of welcome from the GOC Hong Kong. YMCA ladies dispensed tea and sandwiches. It was a change from American coffee and doughnuts in Pusan and Seoul.

I saw a newspaper kiosk on the quayside and bought an air-mail edition of the *Manchester Guardian*—the first I had seen in many months. In the Middle East, there were clashes between the Egyptians and the British at Port Said and Suez. I did not know that I would be there myself within six weeks.

A train took us out to the New Territories, where we were grouped with the Argylls near the frontier with Communist China. Our camp had the beautiful name of Fan Gardens Camp, at Fan Ling. It was comfortable, the billets and officers' quarters a superior type of Nissen hut, with windows to let in the sunlight which bathed the landscape. The officers' mess was a private villa, requisitioned (as we later heard) from some rich Englishman with a good eye for a beautiful landscape. That was before the days of the Long March and the sealing of the border a few miles away. The Englishman had departed and the army inherited not only the cool verandahs, the fine lacquer tables and screens, the polished floors of the reception rooms, but also the exquisite gardens. Lawns curved away between exotic trees and bushes, glowing with colour. Beyond the falling landscape, carefully devised from the perspective of the verandah, brown hills in the Chinese Peoples' Republic completed the frame.

But there was work to do, mostly in the form of guard duties at the frontier. These were frequent and very taxing. The guards were posted in various positions on top of nearby hills, and our area covered a tract called 'Dockwell's Ridge'. A duty officer and several guard-posts stayed there for two or three days at a time under conditions not very different from Korea. The days were hot, and look-outs had to spend most of them peering through binoculars at the shimmering hills beyond the frontier where the Chinese had built elaborate defence works. We observed them: they observed us. It was tedious work, with no activity to break the monotony.

For the officer on guard duty, the return to the mess at Fan Gardens was an extraordinary luxury. One took a leisurely shower to get rid of the dust from the weapons trenches, then to the mess to sit on the verandah and call for a chilled lager at noon. Then lunch, usually cold cuts, sometimes a curry, and after it plenty of fresh fruit from the orchard behind the villa.

I discovered a hand-wound gramophone and some records at the brigade educational office. Mozart symphonies, some Beethoven, Dvorak and Brahms. I found a stretch of lawn near the officers' quarters and lay on my back under the sun. In the mess, a number of books had accumulated from successive regiments, and the eccentricities of taste which regularly crop up in British army units provided some reading. Apart from Waugh, Aldous Huxley and Graham Greene, there was plenty of Liddell-Hart, and a copy of Jomini's *Précis de l'Art de la Guerre*. There were even some books which the owner of the villa—or perhaps his wife—had bequeathed to the military. One which struck me particularly gave the names of flowers in translation from the Chinese. I stretched out on the lawn in the sun, with the scent of the flowers drifting across, and read the names the Chinese had given to their flowers. A white chrysanthemum with a blush of pink at the tips was translated as 'evening sun on a white duck's back'. Others were 'yellow orioles in the green willow'; 'golden phoenix holding a pearl in its mouth'; 'honey-linked bracelet'; 'spring swallows in an apricot orchard'; 'white crane sleeping in the snow'. I scribbled one or two poems and sent them to KS in London.

A meeting was called in the officers' mess to discuss regulations for mess dress, arrangements for guest nights, dining out, and other matters of protocol for maintaining good relations with the Hong Kong community. It was an understandable attempt to recapture the life of a regiment in a foreign posting where ordinary civilised conditions existed. As the evenings were still warm, the CO ruled that 'monkey jackets', wing-collar and black tie would apply for guest nights. When the colder weather approached, number one dress or patrols would be the rule. We were instructed to have these sent out from England together with service dress and other kit left behind. For plain clothes, a white linen suit would be the rule.

On guest nights, some of the special ritual of the Royal Ulster Rifles was restored. Regimental custom involved a piper, and on special occasions three pipers, marching round the tables during dessert as the port circulated. It was ear-splitting in the confined space of our present dining room. At the end of their performance, the pipers were each handed a glass of Irish whiskey which custom again required them to toss back in one gulp. Some wild Gaelic shout preceded the downing of the whiskey.

We dined well, as there was no shortage of excellent Chinese cooks in the kitchens. Sunday lunch, when no duties awaited in the afternoon, was usually a curry of immense extravagance in its trimmings and accompaniments. Apart from huge quantities of molten curry and mountains of rice, there were endless dishes of nuts, sultanas, onion rings, diced pineapple, shredded coconut, sliced bananas; indeed, anything the president of the mess was asked to include to make the curry lunch to end all curry lunches. The table groaned under the various recommendations from officers who had eaten exotically in different parts of India or Ceylon, or perhaps restaurants in Turkestan or Kashmir. Topping off our loaded plates, to sweeten and cool the steaming volcanic mound, we added dollops of imported Indian chutney. The bottles were carefully checked for the authentic label before the officers trowelled it on to their plates. Around 2.30 in the afternoon, swollen and surfeited, the mess would totter off to its quarters, or unbutton and lie supine under the sun, belching between dreams of serving girls and red-hot curries.

Hong Kong and Kowloon were about forty miles away: almost too far for a casual evening out, but there was a reasonably good train service and most of the men spent their evenings there when they were not on guard duty. Apart from the hours on duty, we saw much less of the men now: the separate world of officers' quarters and the troops' billets brought little contact. When four or five subalterns decided to go into Hong Kong, they hired a taxi. The bar and lounges of the Peninsular Hotel at Kowloon was a favourite place for meeting European women. Air hostesses stayed there between flights, and one of the RUR company commanders (a married man, but then this was the Far East) was at the bar almost every evening in service dress and Sam Browne belt, the small crown on his epaulettes conveying his rank. Probably he felt this maximised his opportunities. He was not the handsomest officer in the regiment, but he tried often enough to edge his way along the bar to unattended females. After one or two repulses he would seek solace in whiskey. Then he tended to go off to the red light district in Kowloon if there was enough liquor in his belly.

At Fan Gardens camp, my window faced his room across the space of a few yards. You knew when he had been on the whiskey and women as his curtains remained drawn after reveille, and around mid-morning extraordinary retching noises would come from his room. Although company commanders normally turned out for roll call parades, the first of the day, he usually missed them. The CSM would wait, with the men drawn up for inspection, facing the direction of the officers' quarters. If a fellow officer had not already tipped the wink, the senior subaltern present would take the inspection after a short interval of silent waiting.

But this company commander was an exception. In fact he was not an Ulster Rifles officer, but had been seconded from another regiment in the Irish Brigade. One could detect, from the gossip that he had dashed his chances of promotion some time ago; that he knew he would never rise above his present rank, and so had ceased to care.

If the ordinary Chinese in the nearby villages liked the British presence they certainly did not show it. Some of the villages contained Communist agents or sympathisers and it was forbidden to go there unaccompanied. By day, travelling through on a landrover or a truck, clenched fists were raised furtively here and there, but it was thought best to ignore them. Occasionally there was hatred in a peasant's stare.

The tailors, shirt-makers and shoemakers who flocked to the army camps seeking custom were glad of the British presence. Silk shirts, linen suits, and shoes were absurdly cheap. The shoemaker placed our stockinged feet on pieces of cardboard and drew a pencil line around each foot. The fitted shoes appeared next day.

In Hong Kong the British troops were on the whole well-behaved, apart from the occasional brawl, or a café owner chasing men who refused to pay a bill. The teeming population provided the men with all the brothels they could wish for, and prophylactics were provided free in small bare offices in Kowloon, open twenty-four hours a day.

CHAPTER 22

Suez

I had written to the colonel of the Border Regiment within a few days of arriving at Hong Kong. I said that now that my tour in Korea was over I should like to return to the Border Regiment. If he could assist this in any way I should be grateful. I had decided to return to the battalion before submitting my resignation. Some sense of residual loyalty or of what constituted proper form must have prompted my action. I did not wish to send in my resignation before returning to the regiment which had, after all, admitted me to their family from Sandhurst. But my mind was now even more firmly made up to leave the army at the end of my five-year engagement.

General Shears wrote back promptly from London. He had had a word with the military secretary at the War Office, and I sent back my thanks. I had already heard that the Border Regiment was leaving Barnard Castle for the Suez Canal Zone.

My course of action after leaving the army was now decided. It had grown in my mind rather as an underground stream wandering beneath the crust of things, finally emerging to the light. I would seek a place at Oxford or Cambridge. I was not certain what I wished to read, but three years of extended thinking and reading seemed to me more important than any specific object of study.

At times the War Office moves with surprising speed. Within a week of General Shears's letter I received a posting order, directing me to return to the Border Regiment, listed as 'en route to MELF' (Middle East Land Forces). Armed with this, I called on the adjutant of the Ulster Rifles, who told me to contact 'Movements' at Kowloon. Things happened quickly. I was given a place on the *Empire Fowey*, leaving Hong Kong a couple of weeks later. I packed my kit, returned the gramophone and records to the Brigade Centre, and took leave of my platoon, then of the officers of the Ulster Rifles.

I did not see the battalion second in command at lunch, but I wanted to say goodbye to him. I told the driver of the land-rover to make a detour to his office. He was at his desk, as I expected he would be. Gerald Rickford was the type of officer which the British army—perhaps any army—probably breeds no more. His first and last loyalties were to his regiment—that is, to the men in the ranks, whom he served in the most literal sense, with total dedication. He was quiet, reserved, never raised his voice in the mess and he had the sort of frank, unwavering eyes which only men of complete integrity possess. He accepted the drudgery of barrack life, with its attention to detail, its end-less parades and inspections of this, that and the other, its pro-cession of interminable disciplinary offences, its monotonous routine, as a job worth doing.

He rose from his desk and came round to shake hands. He was a man of few words, little sentiment, and rarely smiled. He said in matter-of-fact tones: 'We're grateful that you joined us. You've been very loyal.' That was about all. Coming from him, it was much more eloquent and it meant much more than friendly toasts from one or two of the subalterns at lunch.

I boarded the *Empire Fowey* about an hour later. She was a big ship, originally German, and said to be the most comfortable of all the troopships in service. The lounges, smoking rooms, bars and dining rooms were as spacious as a Cunard or P&O ship. Even the troops were well quartered. Officers from each of the three services were among the remaining passengers. There were also a number of officers' families on board, army, navy and RAF. Some were accompanied by husbands, some not. The unattached wives were sailing to rejoin husbands posted earlier to England, or going on ahead to prepare a new home in England for hus-bands who would follow them later. By the time we reached Aden there was much adultery around the ship. The chief stew-ard in the officers' bar told us that it happened on every trip. The long haul across the Indian Ocean, whether going east or west,

was always the crumbling point for women aboard, married or
single. He added a couplet favoured among the stewards in the
first-class accommodation:

What the church calls sin and the law adultery
Is much more common when the weather is sultry.

The ship's wireless gave us news bulletins each day, and there
was a printed newssheet besides. It told of increased trouble in
the Suez Canal Zone between the British and the Arabs. In the
bar we made jocular comments about 'getting our heads down'
as we moved up the canal. Others wanted to get at their weapons
to 'shoot a few wogs'. At Aden we picked up mail, some of it
flown on from Hong Kong. I was called to the ship's adjutant,
who passed a wireless message saying that I was to join the advance party of the Border Regiment at Giniefa, in the Suez Canal
Zone. I was to disembark at Port Said.

'Hard luck,' he said. 'Must be some sort of balls up. You're due
some UK leave after Korea, I should think.'

I did not really mind. For me, as for many on board, the long
voyage had become tedious. The thought of dry land—even in
the Suez Canal Zone—promised immediate change.

The long haul up the Red Sea, with the heat so intense that
sleep became difficult, increased the febrile quality on board. By
now, adulterous couples did not trouble to hide their liaisons.
As the night approached they left the bar arm in arm, heading
for the upper deck, where there were dark corners and canvas
mattresses piled near the bulkheads. If you could not sleep during the steaming nights, an evening walk under the stars on the
top deck was a sort of obstacle course, unless you kept to the
middle of the open deck. Some of the unattached women had
single cabins, of course, and we knew which of the subalterns
had struck gold. There were no secrets after Aden.

It was mid-December, and there was fog in the Bitter Lakes.
We moved slowly, the shore line glimpsed now and then like a
khaki mirage suspended in the air. There was a queue of ships
for the final stretch of the canal at Lake Timsah, and the *Empire
Fowey* took on a pilot late in the evening. We moved up the

narrow canal at about five knots. It was night time and the banks of the canal on either side were spectral apparitions bathed by the desert moon. I stood on the upper deck and heard the music coming from the lounge, where another fancy dress ball was under way. At dinner, my neighbours commiserated about my disembarkation. A commander in the Royal Navy bought me a farewell drink in the bar after dinner. 'Port Said,' he said contemptuously, 'the arse-hole of the world.'

I got off with the pilot who had guided the ship up the last narrow stretch of the canal. It was after midnight, and some of the revellers came to the gangway as I stepped off. A small group of friendly faces appeared suddenly in a knot. A bottle of whisky was thrust forward, ornately wrapped. Then I clambered down into the black night to the launch bobbing below, clutching my whisky against my side. I supposed that my jungle green would look rather incongruous against the desert sand by day.

The launch took me over to the Port Said transit camp, with my canvas holdall and few belongings. The transit camp was silent, seemingly deserted, but there were strict security checks at the gates. I was dropped off at a sagging tent near the perimeter. The barbed wire fence was high, like a prison. An orderly corporal, disgruntled to be roused from his bed, brought me a couple of sheets and some evil-smelling blankets. He carried a torch. There were no lights in the tent. Now and then stray shots rang out in the night.

'The Gypos,' he said perfunctorily. 'They keep sending the odd shot across the wire to wake you up. No lights allowed after Lights Out. A bit of trouble in Port Said the last few days. Out of bounds now.' He vanished, with a lazy salute.

I tried to sleep, though it was difficult. An occasional shot cracked overhead. The range was no more than a hundred yards, judged by the crack and thump rule-of-thumb. But the trajectories were high, mostly well above the height of the tents. One low shot brought a quick scuffing across the canvas. Next morning I noticed the pepper pot holes in the roof of the tent and wondered how many nights I might have to spend here.

At the camp commander's office I heard that a few subalterns from the Border Regiment were already in the camp. He indicated where they were on a wall plan and when I asked to move there he agreed. He checked where I had spent the night.

'Christ!' he muttered. 'You shouldn't have been put there. That's next to the wire. None of those tents are occupied. Who put you there? That bloody corporal again. He's trying to get posted home to his wife and kids. He'll try anything to work his ticket. I'll have him by the short hairs. We lost a sergeant there a couple of weeks ago. The wogs operate at night round here. For God's sake watch your kit. We can't trust any of the servants.'

The subalterns forming the advance party of the Border Regiment were helping to unload a ship in Port Said harbour, with a detachment of men. I found them aboard the SS *Chilka*, an ugly cargo ship. They were in khaki drill and did not recognise the jungle green figure waving up at them from the lighter. First came disbelief, then recognition.

'Christ almighty,' was the overture. They thought me mad to have got off the *Empire Fowey*. You could always blame some mix-up in posting orders, say you had to get home for your kit. But it was too late to do anything about this now. I joined them, helping to discharge the cargo from the *Chilka*, and spent the rest of the day unloading stores on to the lighter below. We continued next day. I began to wish I had gone on to Southampton.

I thought of England again as I tried to stomach the appalling food at the transit camp. Apparently they made it deliberately bad so that no one would hang around for a day or an hour more than they need. Even the slovenly orderly corporal's attempt to work his ticket home began to make sense. The permanent staff at the camp were confined to barracks because of the troubles in Port Said. British soldiers had been stabbed. In December 1951 the Port Said transit camp may well have been the worst posting in the world.

Fortunately, we had only two more days of unloading to do, then we left for Giniefa as the advance party of the Border Regiment. Unloading the *Chilka* was a tedious job, but it pro-

vided two vignettes of military life. I was in charge of ten men
at the forward hold, with three or four of them down in the ship's
bowels, the others on the barge by the ship's side, stacking crates
of food or provisions slung down to them. On the second day we
unloaded hundreds of crates of canned beer. The men working
in the hold managed to drop a load or two, expertly, so as to
split the wooden crates. They shouted their apologies up to me
as they placed the cans to one side. They vanished during the
day.

As the morning wore on I was vaguely aware of being one
man short on the barge below. It was difficult to keep exact
count as they scurried about the crates, stacking them carefully,
shoulder height, then head height. But some guilty movements
and furtive looks among the men below brought me down to the
barge, probably more suddenly than they expected. The short
tunnel of crates faced away from the ship's side. Inside, by careful construction, they had made a neat little room. In the dim
light, the two split crates and the empty cans told the story. I
had in fact been one man short, but one at a time. Two spars
of wood held up the crated roof. I swore a good deal, told them
to seal up the cave, and said that if there was any more of it
they would all be in the guardroom on charges of stealing army
stores. I left it at that. They were on a tedious, exhausting job,
which would normally have been done by the local Egyptian
labour force but for the recent troubles. They took it in good
part, except for the one who had clearly not yet had his turn in
the bar.

Next day brought incident number two. The cargo was sugar
and flour in hessian sacks, so there were no stealing problems.
But during the day I noticed a small rowing boat, or 'bum boat'
as the military called them, with its Arab oarsman, approaching
the stern of the ship. The boat had one passenger, a man in a
suit, with a satchel under his arm and a red Egyptian fez on his
head. My suspicions were again aroused, so I hid behind a bulk-
head where I could watch. The man in the fez came nimbly up
the ladder at the stern of the ship, and I was about to challenge
him when I saw one of our corporals already at the head of the

P

ladder, apparently expecting the man's arrival. Both of them looked around furtively. Neither thought to look aloft where I was positioned. The man in the fez reached inside his satchel and handed over a folder to the corporal, who scrutinised the contents, then reached inside his own pocket to hand over something. The Egyptian tucked it quickly into his inside pocket, then went down the ladder again.

I came down to the lower deck quickly. The corporal looked startled as I came up. I pointed to his pocket.

'I'll see what you've got there, corporal. I was watching you.' It was unpleasant to think that I might have to make a serious charge against the corporal, perhaps of passing military secrets, or details of troop movements to the Egyptians, in exchange for a substantial sum of money. The corporal flushed and passed the folder over to me. I opened it to count the money he had been paid. It was not money, but an assorted collection of pornographic photos. The *leitmotif* was oral sex. Most of the photographs were tattered, as though they had been on the market for some years. Some of the characters had a distinctly thirties appearance. One showed a Bertie Wooster figure in dinner jacket, much unbuttoned, smoking a cigar whilst the woman knelt at his loins below. I handed them back, since he had paid for them.

'Get rid of them, corporal.'

'Yessir.'

I walked off. I was not sure which section of the Army Act covered this offence. Probably the section relating to 'good order and military discipline'. But there were times when it was best not to think too legalistically.

CHAPTER 23

Trying to Resign

We had orders to go down to Giniefa, near Fayid on the Great Bitter Lake, to take over the camp where the Border Regiment would be located on arrival. The regiment had now sailed from England with other elements of the Thirty-Ninth Brigade from Dover and Folkestone. Two subalterns and myself travelled down in the landrover laid on by the RASC at Port Said. In Ismailia we lost our way and had a narrow escape. We were driving along a main residential street leading south to connect with the main road down the canal, but saw that it was sealed off by a barricade as we approached the end. We noticed the caps of British military police bobbing about behind the apron of wire. They seemed very agitated. The officer in charge came up to us on the other side of the barricade.

'Do you realise you've just driven along Bomb Alley?' The name struck a chord. All the airmail editions of the newspapers had been talking of the street in Ismailia which the British had sealed off because of bombs regularly lobbed from upper stories on to British army vehicles, as well as continuous sniper fire from concealed positions. We apologised and said we had no map. Our driver pushed back his beret and wiped his brow. The military police gave us directions out of Ismailia and we went safely through Fayid, down to Giniefa.

Our new location was a collection of Nissen huts on stony desert by the side of the road. There was no furniture. What was called the mess was a Nissen hut with a partition across the middle, pierced by a plywood door. Some stores arrived and we unpacked them. We ate off trestle tables and sat on packing cases. We unpacked some tents and moved into them, one each. We shaved in a mug of water and walked off into the wilderness when our bowels required it. The nights were getting cold, and there was no drink. On Christmas evening I sat shivering on a packing case in my little tent, a blanket wrapped around me, reading by a paraffin lamp. So far, the magic of desert nights was absent.

Things improved as more stores were unpacked. The battalion arrived a week later by troopship. It was good to see a number of old friends again. They were in khaki drill for daytime wear, whereas I possessed only the jungle green uniform for the Far East theatre. I had a word with the adjutant, who brought up the idea of getting me on a course in the UK, where I could get my kit together before rejoining the battalion.

I felt I had to mention that I intended to resign my commission and leave the army if possible before the end of my five-year engagement, which would come in just over a year's time. The adjutant was an old friend and took a fatalistic view.

'They may not let you resign. We need a platoon commander for the Anti-Tank Platoon in Support Company. It has to be a regular subaltern. It means a course at Netheravon in Wiltshire. We can get you on a plane pretty quickly if you'd like to take it on. Next course begins next week. Best take it, there's no leave from here just now.'

GHQ at Fayid found me a seat on a military plane within a couple of days. I snatched a couple of days at home before the course began, collected my kit, then assembled with about thirty other subalterns in the dank English winter at Netheravon.

Our weapon was the seventeen pounder anti-tank gun, a cumbrous, accurate, long-snouted piece of ordnance with a high velocity and good armour-piercing capacity. For eight weeks we studied its ballistic capabilities, manhandled it in the mud, stripped the breech, loaded it with dummy shells, aimed it, swung it about, and finally fired it live by day and by night. The course was thorough and well taught, with the accent on practical demonstration, then repeated practice for the participants, and finally the task of giving instruction oneself. By the end of the course we had all mastered the gun, and could teach others to handle and fire it.

But my interest was marginal. In the third week of January I had drafted my resignation to the colonel of the regiment. I kept it for a few days, then sent it to General Shears at the end of the month. I knew now that I would need his help to persuade the War Office to accept my resignation. I had already discovered that it would be resisted strongly. The army had spent—and was spending—a good deal of money on making me a proficient officer. Perhaps understandably they were reluctant to sacrifice their investment. My letter to General Shears tried to convey tactfully that an unwilling officer could not be a good officer in the long run. His reply was more kindly and understanding than I had expected. He asked me to think it over carefully, perhaps for a few months, and, if I felt absolutely certain that this was not just the occasional spasm of boredom which most young officers experienced, then he would help at the War Office.

I had kept in touch with my friend and contemporary from basic training days at Holywood Barracks in Ulster, Dick McWatters, who was now up at Trinity, Oxford. Before returning to the Middle East I paid him a visit. He met me at the station and we walked round Christ Church Meadow. His father, Sir Arthur McWatters, was secretary to the University Chest at the time and they lived in the Judges Lodgings at 16, St Giles, a tall house behind iron railings just north of St John's College. We ate toast before a big fire, then listened to records of bird songs.

Before the light began to fade Dick took me up to the roof of the tall house, where a flat expanse high above the surrounding houses gives a rare view of Oxford's skyline. It was a day of fitful sunshine, with a March wind bringing clarity to the air. But now in the early evening the wind had dropped, and pale sunlight lingered on the cupolas and towers. Looking about, I saw the magic mixture of Gothic and Palladian. St Mary's spire reached high by the dome of the Radcliffe Camera. Beyond, the slim pencil of Magdalen Tower and the bluff crag of Merton Tower to the south. Between them lay the buttresses and crenellations of the colleges and halls. At that moment the die was cast. If I should have any choice in the matter, Oxford it would be.

General Shears advised that my letter of resignation should go to the adjutant general at the War Office. I thought it best to discover whether the army would indeed release me, and sent in my letter of resignation from Wiltshire towards the end of my course there. A week later I was summoned to the War Office. I thought that some much decorated general would interview me. I assumed that adjutant generals are rare creatures. This was not to be. A mild, middle-aged man in a suit met me in the corridor. He looked an archetypal civil servant, with my file under his arm as he sat me down in a small bare interview room. A cup of tea arrived for him, none for myself. He said that he was the military secretary, or a member of that department, and asked about my reasons for resigning.

I mentioned my intention, or wish, to go up to university. Had I got a place? Not yet. A pause. Would it not be wiser to gain a place first before resigning a regular commission? Not really, as my intention was firm, whether or not a university place resulted. Well then, he ought to make it clear that my present posting to Middle East Land Forces made it very difficult to agree to my resignation. There was an emergency situation and Suez might well be a theatre of war if things got worse. Perhaps I should think it over for a month or two, when the situation might be clearer all round. The army would be very reluctant to lose me. And so on. I thought that at this stage it would be wise not to offend the authorities, and agreed to his suggestion. It was also agreed that I could re-submit my resignation after 'a few months'.

I flew back to Fayid. The Border Regiment had moved to Tel el Kebir, about forty miles west of Ismailia, along the road to Zagazig. I found them at Falaise Camp, a barren, treeless site not far from the Sweet Water Canal, a stagnant stretch of water between Zagazig and Ismailia, its name presumably intended as a bad joke. Falaise Camp was little more than a series of billets set on stony desert. Water was in short supply. Flies and mosquitoes were a constant hazard. The food was abominable, hardly any of it fresh, except for the big watermelons bought locally. These quenched the thirst, but there was no nourishment in them. The battalion

had done what it could to make the place habitable, but it was
obvious to everyone that Tel el Kebir was—in the common ver-
nacular—'a shit-awful station'. The effects of it upon the officers
and men began to show as the months passed and the full heat of
the desert sun came into its own.

The Suez Canal Zone was now thickly populated with British
troops and military installations. What little remained of Arab
communities outside the main towns of Ismailia, Port Said, or
Suez were mere dormitories, providing cheap labour for the
occupying troops. Apart from this, we saw practically nothing of
the Arabs. As in Hong Kong, the local entrepreneurs who did
well out of the occupying forces gave the illusion of welcome. The
hairdressers, the shoemakers, the shopkeepers in Ismailia, the sly
men who sold Swiss watches from small kiosks, all smiled and
fawned. Their private thoughts might be different; it was impossible to tell.

The Border Regiment was brigaded with the Royal Inniskillings and the Buffs to form the Thirty-Ninth Infantry Brigade,
part of the Third Infantry Division, with headquarters at Fayid.
Our duties at Tel el Kebir were largely guarding stores. Falaise
Camp lay just outside the perimeter of the vast ordnance and
supply depot we had to guard. Several high fences formed the
perimeter, which was several miles in diameter, with mines sown
thickly between the barbed wire emplacements. Life was slightly
more comfortable inside TEKBLOC, as the vast depot was called.
Inside there were even one or two gnarled and stunted trees to
give some relief to the landscape of barren billets.

Guard duties, which came frequently, involved taking a detachment inside TEKBLOC to brigade headquarters. There, the
officer on duty remained awake by a field telephone set for his
twelve-hour night spell, with visits to the various guards posted
strategically around the inside perimeter wires. The positions of
the guards were changed frequently so as to confuse the Arabs in
the small town of Tel el Kebir nearby. Organised theft from the
vast assortment of stores was clearly one of the mainstays of the
local economy. Hundreds of local Egyptians were employed by
day in TEKBLOC, mostly on unskilled, menial tasks, and every

employee was frisked at the gates when he left. The Arabs' volum-
inous ankle length shirts made this a tricky and a not very pleasant task for the British soldiers who had to do it. Like poor thieves
anywhere, the peasants went to great and often ingenious lengths
to conceal things on their person. The paradigm case, quoted to
us as a warning of their ingenuity and determination, was the
employee who managed to smuggle expensive, high performance
radio valves out of the gate by suspending them from his penis.

The Arabs in the neighbouring villages also made regular
attempts to creep through the perimeter wires during the night,
using wire-cutters stolen from TEKBLOC. Some of them were
probably TEKBLOC employees by day, and so knew how difficult
it was for us to patrol such a huge perimeter, despite the sweep
of searchlights mounted at intervals on towers. Occasionally they
blew themselves up on a mine, but this was very rare, considering
the frequency of the attempted break-ins. The inner wire fences
showed the effects. They had clearly perfected some means of
feeling their way forward through a minefield, with delicate
fingers to detect where desert sand was firm, and where it had
been disturbed. More rarely still, our guards caught sight of the
intruders, and one or two wretched Egyptians would be cut to
pieces by machine-gun fire under the searchlight beams.

I took over the Anti-Tank Platoon in the Support Company of
the Border Regiment. As yet the platoon had no guns, but I
arranged to draw six guns from the ordnance stores in TEKBLOC.
It was an odd experience, 'drawing' six large guns, six carriers to
tow them, then signing for all the bits and pieces which went with
them. We clanked across the stony earth to Falaise Camp, raising
a small dust storm. I lined them up outside the company office
and the company commander looked them over.

'Don't lose one,' he said cheerfully, 'or you'll lose your balls at
GHQ.'

Some of the national servicemen in the platoon were terrified
of the gun. When the day came to take the guns up to El Qantara
to fire them on the anti-tank range there, one small Liverpool
man looked very scared and I expected trouble. I arranged for

him to fire last, but as he had to take his turn as gun crew, handing forward the shells, this did not solve the problem. The blast
of the gun was always dangerous for the ear drums, and the long
recoil, with smoke curling from the breach like dragon's breath,
upset some of the men. The Liverpudlian burst into tears and
closed his eyes as he pulled the firing handle. His beret sailed up
in the air and vanished into the breech as the gun recoiled. We
winched it back and he stared incredulously at the tattered beret.
Then he put it on, with a sudden gleam of pride, and strutted
around with his trophy. After that, we had no more trouble with
him.

Back in the mess at Falaise Camp, the officers tried to create a
semblance of civilised living, but the complete lack of amenities
made it an impossible job. The mess itself was a crude building,
with shutters to keep out the daytime heat, but with a corrugated
iron roof—just the thing to transform it into an oven under the
desert sun. Showers were rationed for everyone, and, despite rigid
regulations on hygiene, the fly and mosquito menace grew with
the approach of summer. Mess dress was monkey jacket and blue
patrol trousers. On some nights the mosquitoes arrived like a
plague and swarmed cunningly in the darkness beneath the dining
table. Our ankles were eaten alive. Some officers gave up dinner
and sprinted to their beds and mosquito nets. Another health
hazard was desert sores, gargantuan boils which, when they
broke, left craters as big as a penny and as deep as a sixpence in
the calf or the thigh.

There was no mains sewerage in the camp, and latrines were
always a problem. We sank deep pits and laid beams across the
chasms, then ran hessian around the construction. Problems arose
when the latrine had to be filled in and re-sited. The wooden
beams had to be used again, but they tended to be deeply encrusted with faeces, with layer upon layer of coarse army toilet
paper glued to the timber. It was the least popular fatigue among
the men.

A simple construction was used for the urinals. It consisted of
no more than a length of pipe, about the same bore as a suburban

drainpipe, forced into the sand at an angle until it protruded to
the height of the average man's thighs. Two or three pipes might
be lined up in a file, at different heights to suit different men.
Some urinals were shielded by hessian, others were not, and NCOs
anxious for promotion (or perhaps they merely had a perverse
sense of humour) occasionally gave a squad an 'Eyes Right' as
junior officers made water at the pipe. Whether to return the
salute with the right hand, holding one's member with the left,
was a question of personal *sang-froid*. Most officers thought it best
to ignore the compliment and concentrate on the matter in hand.
A Part I Order from the adjutant on company notice boards took
care of the matter.

'Warrant Officers and NCOs will not pay compliments to officers when passing urinals.' The hazard ceased.

The normal consolations of a foreign posting were absent. There
were no women in Falaise Camp and, according to a subaltern's
reconnaisance party, none in TEKBLOC either, except for one or
two nurses and WRACs whose favours were already pre-empted.
We were beleaguered in an arid stretch of desert guarding a
gigantic supply and arms dump half way between Ismailia and
the Nile Delta. The local villages were out of bounds because of
the worsening political situation, and in any case the abject poverty in the mud hovels would have made female company there
an immediate health hazard, even if the Arab menfolk had not
closely guarded their thin, undernourished daughters.

After a couple of months, one came to accept Tel el Kebir for
what it was: one of the worst foreign postings in the world. It was
intriguing, partly saddening, to see the effects on newly arrived
officers as this truth began to tell. Among our newly arrived
officers was a rotund, amiable, idle captain, of independent means,
who had finally been edged out of some sedentary posting in
Britain to rejoin the regiment at Tel el Kebir. He liked good
food, good wine, and enjoyed the temptations of the flesh when
these were available. He stood with a whisky in his hand after his
first dinner in the mess at Tel el Kebir, somewhat crestfallen, the
food having plunged him into despondency. He asked whether

the locality did not at least offer other forms of satisfaction and addressed the subalterns sitting stirring their coffee. He had a nice turn of phrase.

'Anyone know a place round here where one can put the meat, then?' The answer was no, unless he wanted to risk a pox in Ismailia, thirty miles away.

He looked glum, and called for another whisky. There was something of the philosopher and the good-natured fatalist in him, however, and he accepted the common lot. Most regiments have, or had, one of him in their midst. Indolent, slightly eccentric, a private income relieving him of ambition, he was liked by his fellow-officers, by the NCOs, and also by the men, who regarded him as a sort of mascot. His military deficiencies had to be concealed for the good of the company and sometimes for the honour of the regiment. But the incipient philosopher in him could fetch up a pithy, telling commentary on military life or on our present predicament. One evening he returned to the mess after completing his rounds as orderly officer of the day. His whisky safely lodged in his hand, he pronounced a neat summary of life for the ordinary soldiers at Falaise Camp.

'Poor buggers,' he observed. 'They come off twelve hours' guard duties; they get their sausage and mash in the cook-house; they stagger over to the NAAFI canteen to fill up with bangers and beans; they down five bottles of warm beer; they stagger across to their tents, climb under their mosquito nets, have a bloody good wank, and drop off to sleep. End of a perfect bloody day.'

To relieve the tedium a generous use of army transport was allowed to ferry the men into Ismailia when they were not on guard duties. The officers also went into Ismailia, sometimes with the troops, more often in a landrover. There was a reasonably good NAAFI club for the men, a YMCA club, and some bars still in bounds, as well as cinemas.

The officers used the three or four sailing and beach clubs sprinkled around the shores of Lake Timsah. The French Club was favourite. It was built for the French colony of technicians, engineers, and canal pilots who serviced the canal and its complex

installations. British officers were allowed to join though the
language barrier kept national groups apart. The French had no
intention of making the further concession of speaking English.
Very few British officers could manage French, so there was little
rapport.

There was also a Greek Club, where the food was worse, but
the music and cabaret more interesting and exotic. Nearby there
were beaches where one could lie under the sun between swimming in the slightly brackish water of Lake Timsah. On most
days, ships would be moving slowly across the lake as they passed
between the various reaches of the canal, and small sailing boats
from the clubs would be dotted about the lake. There was a time-
lessness about life here.

I had brought some books out from England at the end of the
course in Wiltshire. I was not certain what I should be reading to
prepare myself for possible university entrance examinations at
Oxford or Cambridge, but past enjoyment took me to Macaulay's
history again. The rich vocabulary, the sustained antithetical
sentences, moving towards their leisurely predicates, had just the
right self-assurance to recapture the permanence of England.

I submitted my resignation once more, this time through the
CO of the battalion. We talked, and he said that, if my mind was
made up, he would forward my letter to the brigade commander,
and through him to divisional headquarters. He warned me that
I might have to see both the brigade commander and even
General Sir Hugh Stockwell, now GOC of the Third Infantry
Division. Getting out of the army began to look far more com-
plicated than I had expected.

A few days later I was called to brigade headquarters for 0800
hours. The brigade major saw me first, and managed to convey
that the brigadier was very angry with my letter. I thought it best
not to ask why, and in any case there was no time. The interview
got off to a very bad start. The brigade major flung open the door,
said 'Quick March', and I went in briskly, feeling like a prisoner
on a charge. The door closed and I was alone with the brigadier.
I wondered if a charge sheet was about to be read out. I was
standing to attention. The brigadier was writing at his desk,

rather pointedly I thought. He gestured with his pen as he continued to write. No doubt he meant me to stand in front of his
desk, but I misinterpreted the gesture. An easy chair stood before
his desk, and, thinking him to be busy for a few moments, I sat
down gratefully. The nib scratched a few more lines, then the
brigadier looked up. His eyes popped wide.

'Stand up!' he said crisply. 'Who told you to sit down?'

The interview went badly. He inquired about my plans should
my resignation be accepted. University? What on earth had put
that idea into my head? It was too complicated to discuss, so I
was brief, perhaps too brief. He slapped his desk as the interview
ended. He pressed a buzzer and the door opened again. I marched
out.

I became gloomy as the weeks passed. Then a letter arrived
from the War Office in London, signed by the military secretary,
this time a lieutenant-general. The signature did not matter. The
brief, first paragraph gave the good news:

'I am directed to inform you that your voluntary resignation
has been approved in December 1952, on the assumption that
there will be no change in the present policy.'

Six months to wait. But at least my resignation had been
accepted in principle. There was now no possibility of gaining a
university place for autumn 1952, but I did not mind a year's
delay now that I was reasonably certain I would be allowed to
resign my commission. The worst that could happen would be a
further delay, or a war emergency in the Middle East.

I had already written to several Oxford colleges—Corpus,
Merton, University, St John's, Christ Church. The replies were
encouraging. My school record seemed to have been well received, though one or two of the senior tutors said they were
concerned about the four years' absence from formal study. My
languages would be particularly neglected, they surmised, unless
I had kept them up. Perhaps I should arrange to take refresher
courses.

In fact my French was fairly fluent, as it had been one of my
strong subjects at school, and I had kept up my reading ever since.
Moreover, I had met one or two of the French families at Ismailia

and conversation presented no problems. But my Latin was very rusty. An inquiry to the Royal Army Educational Corps at divisional headquarters unearthed a tutor, a national service classicist, who agreed to give me some spare-time coaching when our separate schedules allowed it. We discovered that the only possible regular time was a Sunday morning, so thereafter my Sundays started early, leaving for Fayid at 0800 hours for a weekly three hours with my Latin tutor.

The battalion MT officer was incredulous when I requested a landrover each Sunday morning for Latin studies. At first he took it to be a joke, or a strategem for a weekly visit to a mistress in Ismailia. But I produced the written evidence and he met the request. There was some raillery in the mess when the news got around, but subalterns were quick to find reasons to join my landrover for a day at Fayid, where there were plenty of bars and restaurants.

Each Sunday Captain Ian Thompson, RAEC, sat with me in a hot tent near Moascar going through various texts. I would send him my exercises at mid-week. After a few weeks, when some of the rustiness had been dispelled, the pace became more leisurely, and I was enrolled in a correspondence course in England, again under the auspices of the Royal Army Education Corps. My construes came back by post through divisional headquarters, with marginalia in a small, scholarly hand. I did not know the identity of my tutor, and took it that it was not for me to inquire, but now and then he added a congenial note at the end of a piece of work, and I would send back my thanks. I guessed that my tutor was a retired classicist living in or near Oxford, judging by the postmark on the envelopes, but I never discovered his identity.

I was none the worse for these toils in the desert, since life at Tel el Kebir did nothing to stimulate the mind. The dreary round of guard duties continued, punctuated by exercises and manoeuvres. These had the advantage of taking us out of Falaise Camp to sleep in the desert, and for several nights one had the almost mystical experience of lying under the desert sky, awash with stars. For the first time I felt the curious tug the desert is supposed to exert on travellers; with what seemed limitless open space above and about you, giving a sense of the numinous. Some lines from Flecker's 'Golden Journey to Samarkand' came to mind. The desert nights gave them a special resonance:

> When those long caravans that cross the plain
> With dauntless feet and sound of silver bells
> Put no more forth for glory or for gain
> Take no more solace from the palm-girt wells.

On manoeuvres, however, there were no palm-girt wells. With my anti-tank guns I was usually well forward—sometimes forward of the infantry. On one exercise, Exercise Sandcastle, involving the whole division, the Border Regiment was the forward echelon of the Thirty-Ninth Infantry Brigade, which was itself the forward brigade of the Third Infantry Division, heading west to meet the 'Fantasian' forces advancing from the Nile Delta. I went to various 'O' groups towards dusk, and, as the tactical plan unfolded, I realised that on one flank my platoon was the most forward unit of the division. When zero hour arrived, after dark, I moved forward with my carriers and guns. Putting the whole division on its correct axis and line of advance across open desert now depended entirely upon my compass. With a great deal of hardware behind, and the usual confusion that night-time movements bring, it was not the most enviable position, but at least I had the brief experience of leading a division.

We also learned a good deal about desert warfare, with its special strategies and tactics, and there were moments of exhilaration when orders came through the radio set to go forward, full throttle. The driver of my big Oxford carrier would put his foot down and we raced across the hard, stony plain, six carriers abreast, trying to outpace each other with the guns bouncing behind us and our tracks spinning up a dust storm behind. This again was one of those rare moments when the army recruiting posters came true. It was indeed a man's life; every individual knew his job, and had mastered the technical details required for it. Each crew could uncouple the heavy gun, spin it around, swing out the steel legs to lock them into the earth, fetch up the shells, load and aim and fire—all within sixty seconds.

CHAPTER 24

Cyprus

A relief from the tedium at Tel el Kebir came when I took my platoon on detachment to El Qassasin. This was another supply depot, half way to Ismailia, well off the main road and so even more isolated. The main duties were again guard duties, but detachment at El Qassasin had one special attraction. There was a large and luxurious open-air swimming pool which, for some inexplicable reason known only to high command, had been built at this spot sometime during or after the end of World War II. The depot we were guarding was a Royal Engineers supply base. A possible explanation was that a much larger detachment of the engineers some years earlier had built the pool to bring some comfort to this lonely spot. The cool, clear, filtrated pool stood next to the little camp where we settled.

We shared the small mess of the three or four Royal Engineers officers based permanently at the camp. The OC was Major Upton, a vast, humorous man who had seen his wife and family for only a few months in eleven years through some chapter of accidents in War Office postings. His family were now in England and he hadn't seen them for two years. He grumbled in the familiar self-mocking, ironic manner which preserved an officer's sanity in such conditions. But he had done what he could to make life comfortable for himself and the two or three other officers marooned with him. The most important piece of furniture in the mess was an American-size refrigerator, head-high. Its main function was to keep the beer cool. Pilsner beer came out to the patio ice cold. This special luxury explained an unusual habit among the officers permanently stationed here. At breakfast, which was usually bacon and local Arab eggs (thanks to Major Upton's good relations with the local villagers), the officers drank beer. It was startling, on the first morning, to see a cold lager put down next to an officer at the breakfast table, but we adopted the mess habit.

During the night we had to visit the various guard detachments posted inside the wire perimeter. The usual arrangement of a surprise visit by the duty officer was difficult as the landrover had to approach along an open desert road, then answer a sentry call at the main gate, before passing inside the compound where the main RE stores were kept. This gave the sentry ample time to telephone from his sentry box to the corporal and men in the guardroom at the other side of the depot.

Major Upton was disturbed by the continued disappearance of stores. His theory was that they were being handed through the wire at night. The minor stores had little intrinsic value, and were not much above the level of digging tools, picks, axes, and lengths of metal for constructional purposes; all valuable to the local villagers in one way and another, but the abysmal poverty there made the OC wonder how on earth they could pay a price high enough for our soldiers to risk serious charges.

At the mess we laid a plan, which involved hiding an extra officer in the back of the landrover visiting the guard at night, dropping him off inside the depot, and leaving him to make a sudden call at the guardroom some time after the landrover had left the depot. The plan succeeded. When the officer arrived suddenly at the guardroom, he found the explanation of the missing stores. Young Arab girls were giving their favours to the men in the guardroom. The girls had been handed through the wire by some local entrepreneur. The guard corporal enjoyed special favours—two girls to himself. Apparently it was a regular arrangement. Payment was in the form of army stores.

A serious problem of discipline as well as the theft was involved. Court-martial proceedings would inevitably follow if charges were brought. A difficult decision had to be taken. The losses from stores in the depot were fairly trivial—a dribble of nondescript small items; nothing vital or highly expensive. It would be intensely difficult to fix responsibility, given that payment was in services rather than in cash. These factors, and others, came out as several corporals were questioned separately back in the camp.

Q

The offences had clearly involved successive detachments from several regiments. Our own NCOs were not the instigators of the trade. The contract was regularly renewed by the local whore-master, who took care always to stay outside the wire.

The final decision was for local summary jurisdiction, officially undeclared. The corporal and his men were given extra guard duties. The hole in the perimeter wire was trebly sealed up, and word was passed around, bluntly enough, that any repetition would lead to instant court martial. So the matter ended, unless perhaps the local whoremaster merely awaited a fresh detachment of troops. Meanwhile, the losses of picks, shovels and other minor implements declined sharply. In the officers' quarters some envi-ous remarks were passed. Sometimes the men had all the luck. At Lights Out one night a fellow subaltern went off to his narrow camp bed on a rueful note: 'A pick-handle for a poke. Pretty good.'

We returned to Tel el Kebir for the tedious round of guard duties at TEKBLOC. I made forays to the library at divisional head-quarters at Moascar, where there was also a library of gramo-phone records. Once more I managed to borrow a portable gramo-phone from the Education Corps, and in my room at Falaise Camp I lay beneath the mosquito net, listening to Mozart and Beethoven symphonies on the scratched and hissing discs.

After some months in the Canal Zone, officers and men were allowed a short period of leave in Cyprus. Usually two or three subalterns would take their leave together in order to divide the expense of hiring a car, apart from having congenial company in a new environment. I preferred to go alone, and arranged a time when there was a gap on the chart of optional preferences marked up in the adjutant's office. I wanted to take a good look around Cyprus, above all to be free to head off in any direction I chose.

I was reading a battered Handbook of Cyprus, bought in Isma-ilia and published by Macmillan in 1920. It was written by H. C. Luke and D. J. Jardine. The book's clear but leisured prose contained real scholarship. It traced the history of Cyprus from the time Richard Coeur de Lion had conquered the island before landing in Palestine. There was also plenty of detail on the period when Cyprus was a sanjak of the Ottoman Empire.

I stayed at a hotel in Nicosia looking out over Metaxas Square, at the bottom of Ledra Street. It was here that much of the blood-shed came in the months and years which followed the upsurge of the Enosis movement, but in 1952 the movement was inchoate. Nicosia was merely my base for the exploration I wished to do. Hiring a car was expensive, so a motorcycle seemed the best solu-tion. I haggled with a Turk, who had a sense of humour and a row of gleaming machines outside his shop. We struck a bargain for a week's hire. I went off, feeling as free as the wind.

The island was more beautiful than I had expected. By con-trast with some arid stretches inland between Nicosia and Fama-gusta, the shoreline was a green garden. The era of the package jet tour had not yet arrived, and outside the main towns traffic on the roads was rare; at most a twice a day bus, loaded with villagers, struggling along the littoral.

I discovered hidden villages high among the Pentadaktylos mountains along the northern shore, an Armenian monastery, and then Bellapais monastery. It was startling to discover this Gothic survival under the hillside, its hollow windows staring out to the waters of the Levant. I had bought a camera in Nicosia, and took a snapshot of the village next to the monastery with the men sitting under the Tree of Idleness outside the café bar. Lawrence Durrell and *Bitter Lemons* put it on the tourist map a few years later, and then Bellapais went into the tourist brochures and the village bar bought a jukebox and strip lighting.

The coming political troubles could be detected here and there in Nicosia, where ENOSIS was splashed on the walls around Metaxas Square. One end of Ledra Street was Greek, the other Turkish. Some of the shopkeepers showed in their eyes that the British were not welcome. There had been one or two incidents, mostly at bars and dance halls where the British troops congre-gated. The grenades and the shootings came later. In inland villages Greeks and Turks mixed freely, with no evident strain, and even in Nicosia one could find Greek and Turkish shops next

door to each other, the owners sitting outside in the sun at mid-
morning, sipping their black coffee and glasses of water, gossiping
and fanning themselves. But there was a nervousness in the air,
as though a volcano which had once threatened the community
was now rumbling again.

I stood at the stern of the boat taking me back to Port Said.
The island slipped away until the green fused into turquoise and
blue. Cyprus was a piece of lapis-lazuli, floating on a cobalt sea.
A year later, Greek and Turk were killing each other, sharing
only a wish to get rid of the British.

CHAPTER 25

Christians and Belly Dancing

Back at Tel el Kebir, I again took my turn at guard duties, sitting in the control room at brigade headquarters, trying to keep awake by reading, or by writing letters, or doing my Latin. At Oxford Merton and University Colleges had replied with more encouragement, but said they would like me to sit college entrance examinations early in 1953 if I could get back to England by then. I wrote once more to the military secretary, and a letter arrived some weeks later saying that I would be allowed to resign my commission early in 1953. I guarded the letter like a holy grail.

A week or so later I had a message to call at the adjutant's office. I suspected some disastrous change of mind at the War Office, but the news was quite different. The battalion was asked to supply an officer for a Christian Leadership course at Lake Timsah, near Ismailia. Would I like to go? I reminded the adjutant of my retirement from the army, and added that I was not in fact a true believer in the Christian faith, nor did I think an instructional course under army auspices would lead to conversion. He accepted this, but as he had drawn a blank with other subalterns, who shrank from the prospect of becoming a sort of unofficial padre or regimental evangelist in the eyes of the authorities, everybody was hoping that I would go. After all, in view of my forthcoming retirement, I was the only one in the battalion whom the higher echelons of command would not be able to chase up for religious instruction to the troops. I took the point.

Besides, the adjutant added, glancing over some papers, there was very little to do on the course: the odd lecture, a bit of communal prayer once a day, some discussions, and plenty of time for sailing and swimming at Ismailia. I agreed to go.

We were quartered in tents inside, though slightly apart from,

an army leave centre by the shores of Lake Timsah. Trees flourished in every part of the camp, owing to some ingenious irriga-
tion system. The tempo of our life was very much that of the
adjacent leave centre, with a generous reveille, no parades of any
kind, and an atmosphere of leisure. There were about a dozen
officers on the Christian Leadership course, and after two preliminary talks by a major in the Royal Army Chaplain Corps our
meetings became group discussions. These sessions quickly divided
the sheep from the goats—or rather, those who had volunteered
for the course out of conviction, interest, or faith, and those who
had been persuaded, cajoled, or ordered to come. The majority
seemed to be in the second category. The discussions did not
reach any recondite theological level.

At 6.0 pm each evening we were asked to attend for prayers in
a small wooden chapel under the trees in a quiet part of the camp.
Attending prayers was not obligatory, rather a matter for individual conscience. As we did not like to show brutally that the
Christian Leadership course was failing to produce Christian
leaders, most of us turned up each evening and bent the knee for
the few minutes required.

Afternoons were free of instruction, and perhaps intended for
meditation or devotional reading, but most of the subalterns went
off to swim, play tennis, or sail. I struck up a friendship with a
subaltern in the Royal Sussex Regiment, and we hired bicycles
from the leave centre next door to cycle into Ismailia for lunch
or to do some shopping. In plain clothes, we could merge with
the many civilians—mostly French—living in Ismailia with their
families. My friend knew his way about Ismailia. He mentioned
that although we were not supposed to use a semi-private beach
nearby, called 'French Beach', because it was restricted to French
families, nevertheless it was possible to swim across from a spit of
land, approaching it by water, and so join the company by stealth.
The golden sand was much superior to the grubby, boot-trodden
sand on our side of Lake Timsah. There was also a good bar there
serving chilled wine by the glass. More to the point, he confided,
he had met an attractive French girl on the beach and he believed
she had a friend who lacked company at this moment. There was

an acute shortage of young Frenchmen among the families at Ismailia. Probably the sons were in jobs in France, or completing their studies at university. Meanwhile their sisters languished under the sun, bored and without company, though closely guarded by their mothers because of the dangers for young European women in an Arab city.

I joined my friend's daily trespass and met the girls. Rendezvous were arranged. Thereafter, the four bicycles would head for a pine wood nearby on occasional afternoons. The dalliance was fairly innocent, as the convent-educated girls insisted on being within sight or hearing of each other on the pine needles. But time flew, as it does with lovers, and we had to pedal hard as the sun began to throw long shadows across the littoral. The girls would wave as they turned for home at the Ismailia suburb, and we cycled on, usually arriving late for prayers, so that we had to fling down our bicycles and tiptoe into the hut, bending the knee in the back row, breathing deeply as we prayed for divine help to abjure the temptations of the flesh.

Another letter arrived from the War Office in London, stating that before completion of my five-year engagement in the army I was entitled to take a course of study at the Higher Education Centre at Aldershot. This was presumably some sort of rehabilitation entitlement. An inquiry through divisional HQ revealed that the courses were wide-ranging, from economics and languages to the study of law. It seemed a good opportunity to broaden my studies, as well as get back into the habit of sustained intellectual effort. I replied saying that I would like to take up the offer. As September gave out, so did my time with the Border Regiment and the Christian Leadership course at Lake Timsah.

There was nothing so compromising as passing or failing the Christian Leadership course. The padre in charge of instruction hoped that we had derived some benefit from it, and left it at that. As he shook hands with all of us, I had not the heart to tell him that I would be leaving the army within a few months.

Our brief liaison with the girls from the French beach had also ended, as their mothers came to suspect their regular rambles in the pine woods outside Ismailia. When they failed to appear at our rendezvous we cycled by the apartment building where they lived. A note fluttering down from a balcony told us that they were now confined to the apartment. They waved forlornly from the window, and we did not meet again.

Word came through that I could join one of the study courses at the Higher Education Centre in Aldershot early in October. A posting order followed in a few days. I would sail aboard the *Empire Test* on the old wartime MEDLOC route—across the Mediterranean, up the Adriatic to Trieste, then take the train across Europe to the Hook of Holland.

On the night before my departure from Lake Timsah my regimental colleagues gave a small party. The arrangements escalated as the evening approached—chiefly because the senior officer present, an able and witty major, had sophisticated tastes in entertainment. He hired a three-piece band and two dancing girls from Ismailia. The girls would dance for us in the mess, the major announced at dinner. They might even remove their clothes in order to show belly-dancing to better effect. We doubted that the girls would do anything of the kind: the band leader would probably not approve. In the usual way members of the band would have proprietory rights over the girls. One of the subalterns suggested a plan. The band leader and the girls were told that I was a high-ranking officer, visiting Timsah and anxious to see all the sights. I must not be disappointed in any way. But I pointed out that as we had no field-marshal's or general's uniforms around the mess the band leader would not be fooled. In that case, they said, I would dress in plain clothes, and a fellow subaltern would pose as my *aide de camp*. The plan was approved. The entertainment began after dinner.

The two dancing girls were markedly different in age and appearance. One was young, attractive and shapely. The other— who may well have been her mother—was middle-aged, thin, wrinkled. They took it in turn to dance to the tinny three-piece band. They were accomplished belly dancers, and the effect was to fire the determination of the major and the subalterns who had formulated the plan. I tried to look gruff and displeased as my ADC went up to whisper with the band leader, who threw alarmed glances in my direction, then muttered rapidly to the girls in Arabic. Their movements became more prolix, but there was no stripping, apart from the damask head covers and yashmaks which fluttered to the floor. After some further asides during an interval, it was clear that the older woman was prepared to remove her clothes, but the young one was either too shy or under instructions not to do so. The plan was abandoned.

My colleagues insisted that before leaving the regiment I must show that the army had made a man of me by drinking a vast composite cocktail, made up from every bottle of spirits and liqueur in the bar. The result was a tumbler full of viscous liquid, with the green of the Crème de Menthe defeating most of the paler spirits and liqueurs in the spectrum. I downed it, shivered slightly, and drank a lager quickly in order to drown the taste.

The band packed up not long after midnight. We were slightly alarmed to see the major going off with them. It was not wise for an army officer to go off into the Arab quarter of Ismailia at night. But he knew what he was doing, even though he had had a lot to drink, and reassured us that all would be well. We stayed on in the bar for another hour or more, singing coarse songs. When we finally broke up I made a detour under the stars as I saw the light on in the major's tent. I wondered if he had got back safely from Ismailia.

Like any major who had spent years in foreign stations, he knew the balance of risks, so I was not really surprised to find him safely back. Nevertheless, the picture etched itself sharply on my mind. It seemed to incorporate and to symbolise all the elements of military life in the hot, desert-dry, thankless stations of a shrinking Empire. The harsh electric light glared from the single bulb which swayed from the flex draped across the roof of the tent. Beneath it, the mosquito net descended to the camp bed, slightly askew, with the major's white shirt, blue patrol trousers and shoes scattered upon the single dusty patch of rug. The mosquitoes whined. The major was lying on his back above the sheets, but upside down on the bed, with his feet resting on the white pillow. He was wearing his black silk socks. The rest of him was stark naked. He was sound asleep, breathing deeply, his mouth open beneath the trim moustache. One hand flopped down beneath the canvas bed; the other was resolutely fastened on his penis. I switched off the light and stole away.

I returned to Falaise Camp at Tel el Kebir to carry out the handing over of my anti-tank guns, carriers, stores and other hardware. This had to be done thoroughly, complete with long manifestos and a detailed inspection of guns and carriers. I took care to see that signatures went with every item handed over. I had no wish to receive a bill from the War Office at some future date for a missing anti-tank gun.

Gloom had settled on the battalion and morale was fairly low. Earlier they had fed on rumours that the Third Infantry Division was to go to Germany. Officers and men had conjured images of civilised surroundings, comfortable billets, good food, a more equable climate, and, above all, off-duty hours in cities full of beer cellars, night clubs, and ripe girls with golden hair. The vision had now faded as the political situation grew worse. A rapid deterioration in Anglo-Egyptian relations had developed.

Neither the officers nor the men liked the local Arabs, though this was a consequence of seeing the Arab community at its worst. The poverty, the squalor and the smells of the Arab villages along the road to Ismailia made the barrier between the British forces and the Arab community almost total. Although the British garrisons provided most of the employment for the Arabs living in the Canal Zone, no one could pretend that it was of a type designed to bring economic growth to the region. The Egyptian government clearly had not the slightest intention of injecting the most rudimentary level of investment into the area whilst it was under British military rule. The British, having perhaps read the writing on the wall for longer than they officially admitted, had done nothing to improve the basic economy of the villages surrounding the garrisons in the Canal Zone. As a result, the impoverished Arabs got the worst of all worlds. The roads linking the dozens of garrisons, camps and installations in the zone had been built for the occupying military force. Any usefulness to the Arabs was the merest contingency. As army convoys passed by the villages straddling the main roads, tiny urchins with matted hair and rickets would hold up their hands, palms upwards, like automatons, in case a coin or a bauble was thrown to them out of pity. If none came, the infant hands and arms quickly responded with obscene gestures of unambiguous meaning.

The British lack those qualities of indifference and ruthlessness which occupying forces need if they are to feel easy in the occupying role. The special characteristics of the British—native tolerance, a built-in sense of justice and fair play, a latent sense of humour—could find no purchase in the social and economic conditions in the Canal Zone. As the months passed the eternal guard duties affected morale. The ordinary soldier was fed up and far from home. Married officers, warrant officers and men were separated from their wives and children, who remained in England hoping that their husbands would return soon, or that the rumours of a move to Germany might prove correct. Married quarters in the Canal Zone hardly existed, and the emergency prevented any plans for uniting families.

After ten months, husbands and wives had ceased to feed on rumours. Letters could become less frequent. Frustration became fatalism or despair. Gossip from home would hint at infidelities. Now and then, in the intimacy of regimental life, one heard of a warrant officer or sergeant leaving by military flight for a few days in England. The official, stated reason was 'compassionate leave', but the term often masked marital break-ups and domestic tragedy. The CO and the adjutant would preserve confidences, but county regiments were intimate societies; officers and warrant officers who had served for twenty or thirty years together withheld few secrets from each other. Little would be said; much would be surmised. One did not inquire further, and it was bad form to introduce private concerns into the mess. But it was not difficult to visualise the sorry domestic row in the barely furnished married quarters adjacent to bleak billets at home; the recriminations, the bitterness as a husband tried to put right in three or four days of leave what ten or twelve months' absence had brought about.

The sergeants' mess committee asked me to call in for a drink on my last evening with the regiment. I walked across the camp on a warm night luminous with stars and the high desert moon. The mess was a cosy place—more so than the officers' mess. The bar was like an English pub. All the technical skills in the mess had been unleashed and deployed, together with quantities of army timber, paint and stores, to make it a sophisticated spot. It was strange to see so many of the sergeants and CSMs hatless and in plain clothes. They bought me beer, asked about my future plans, said it must be nice studying at a university, wished they could have had the chance, and drank more beer. We reminisced about minor incidents in the life of the regiment. Some of the CSMs had been in India before the war. I suddenly realised how very raw any new second lieutenant must have seemed to them—whether Sandhurst or national service—on his first day at the company office.

These sunburned, battered men had developed immense shrewdness for their scores of daily decisions in barracks, on the drill square, at defaulters' parades, on manoeuvres, on Adjutant's Orders, on CO's Orders. If they lacked intellectual power, if the life of the mind did not appeal to them, if they were uncomfortable with abstractions, they were nevertheless excellent judges of character. They knew that the new second lieutenant who tried to be chummy with the men, who wanted to muck in, was less a convinced democrat, more an uncertain leader with inner anxieties and a sense of inadequacy: that chumminess produced more problems than it solved; that in war it could prove disastrous if soldiers told their officers to go to hell. The professional experience of these warrant officers and sergeants had taught them that discipline must not be inflexible; that compassion must now and then show through. They knew when to be iron hard with defaulters or trouble-makers; how to distinguish individual viciousness from loss of temper; when to turn a blind eye if an old soldier of the regiment was helped back into camp blind drunk; how to detect instantly the mood of the company commander when he arrived at the company office in the morning—liverish after drinking late in the mess, or in a genial mood, ready to chaff the CSM and the company clerk.

Walking back to my bunk in the hot little hut, I heard the bugler call 'Lights Out' on the other side of the camp. The lonely sound darted here and there among the silver geometry of the billets and company lines.

CHAPTER 26

British Museum

The landrover took me north along the flat road, as straight as a pencil, by the main section of the Suez Canal. The SS *Empire Test* was already berthed at Port Said, so I was spared another visit to the transit camp. The ship was small and squat but this was to be only a five-day voyage across the Mediterranean and up the Adriatic to Trieste. She was carrying about half the normal complement of troops and officers. It was October; the days were warm, evenings cool. The open Mediterranean brought fresh breezes, and on the second evening Crete lay on our starboard side, a long brown huddle of mountains.

On the fourth evening, the bulge of Istria loomed on the starboard side, and we slid into Trieste. I took dinner in a restaurant by the waterfront, intrigued by the clean cobbles, the tramways, the general absence of sand, dust and unmade roads. The rest of the journey was by ordinary train across Austria, Bavaria, and up the Rhine to the Hook of Holland.

When I arrived at Aldershot, the major in charge of the Higher Education Office was at a loss on how to construct suitable courses for me. They were not geared for a Sandhurst-educated subaltern planning to go up to university. The major indicated the resources of the Command Library, hinting that perhaps I might prefer to spend a good deal of my time in private study. Meanwhile I could attend any of the lectures, on any of the courses as I chose. Looking down the lecture list, I saw possibilities for improving my maths and studying law. But the language classes were at an elementary level, so I was thrown back on the Command Library, apart from some lectures on the history of English law.

After a week I switched my studies to London and the British

Museum. For some years I had jotted down on odd scraps of paper the titles of books I felt I ought to read. History, literature, biographies and memoirs provided the bulk. There was no shortage of seats under the dome at the British Museum. The Fulbright programme was not yet in spate. In the Charing Cross Road books were ridiculously cheap. I came across a short memoir which had a profound effect on me. It was written by John Mulgan, a New Zealander who had come up to Oxford in 1933, after three years at Auckland University. He brought with him a love of Greek, but also a mind uncensored by British conventions or society. He read English literature at Merton with Edmund Blunden, took a First, published some poems, and spent the remaining pre-war years working in the Clarendon Press. But it was the war which brought out the real qualities of the man. He entered it in no spirit of patriotism or chauvinism, but with a matter-of-fact recognition that evil must occasionally be countered with force. In 1943 he was parachuted into Greece, and his account, *Report on Experience*, was largely devoted to his work with the Greek resistance movement. But he also wrote of the pre-war years. He died in Cairo in 1945 and the Clarendon Press published his book posthumously.*

I retain the copy I bought in 1952. It has been out of print for some years but this forgotten book seemed to me—and still so seems—one of the most perceptive books to come out of World War II. Perhaps it was a New Zealand boyhood and education which allowed Mulgan to perceive so acutely the underlying reasons for Britain's unpreparedness in 1939. It was not just the amateurism of the War Office, the long weekends in the country; not just that military science was thought to be a distasteful matter, and men like Liddell-Hart at best cranks, at worst dangerous theoreticians: it was also the basic tolerance, the freedom, the plain niceness of British life. Recalling his own undergraduate years at Oxford Mulgan wrote:

> Bad times are said to encourage religion. Certainly we sought for a religion or its substitute. Semi-political opiates like the Oxford Group

*John Mulgan, *Report on Experience*, Oxford University Press, 1947.

were easy to avoid. They were harmless people who neither drank nor smoked, but ate too much and talked about themselves. . . British fascism in its overt form was hardly palatable even for the well-to-do, with its raffish and imitative uniform, its combination of one-time prize fighters and sexual guerrillas. A few earnest patriots like the late Lord Rothermere, looking around for some form of social insurance, took up Mosley and found him too hot to handle.

Looking back to his work with the Clarendon Press in the thirties he wrote:

During most of these years, I was living in Oxford. Chance, rather than ability, set me among academic men whose minds had the perspective of more than one war and more than one generation. I was too young to appreciate then the quality of mind that comes from a disinterested search for truth, either scientific or historical. There were old men whom I met then whose life's work had gone into a Coptic dictionary or the editing of Anglo-Saxon fragments. My connection with them was amateurish and commercial. If I were to go back there now, I should feel a greater respect for them. At the time I resented their aloof and suave disinterestedness, and failed to realise that they, more than most people, were building while the world burned.

The little book seemed to me then, and still seems now, an eloquent and perceptive testimony.

Word came through from the War Office that my retirement was officially granted from early February 1953. There was little more I could gain from the Higher Education Centre at Aldershot. I prepared to leave for the Regimental Depot at Carlisle Castle.

My last afternoon at the centre brought a curiously unforgettable moment in which music fixed the scene with blazing clarity in my mind. It was a cold winter day, dry but with gloomy clouds threatening rain. I was reading in a small study room overlooking an anonymous road at the back of the centre. It was a forgotten stretch of asphalt, marooned by the re-location of billets or army lines. But there was a military cemetery at the end of it. As I sat, the silence was gradually infiltrated by a military band. The noise grew louder until it filled the room. It was a funeral march—the Dead March in Saul, in fact. Looking out of the window I saw a band in full dress, the big drum muffled with a black pall. Just behind was a gun carriage drawn by artillerymen. On it, a coffin, draped with a Union Jack. There were no mourners following. I had no means of telling whether this was a rehearsal or whether a corpse lay beneath the Union Jack on the gun carriage. As the band passed beneath the window, something in the acoustics of the room caught up the dirge and magnified it, so that the sound crashed about the walls, the muffled beat of the drum echoing from wall to wall. The procession vanished around the curve of the short hill, but the music continued, diminishing until it too ceased. Minutes later, it was as though the whole thing had been an apparition. But the knell of the drum and the slow insistent beat of the music stayed in my ears and in the mind as an indelible commentary.

R

CHAPTER 27

Last Post

At Carlisle Castle I met regimental colleagues I had last seen at Barnard Castle before going out to Korea. Depot duty had its frustrations for active officers. In times past there would have been the partial satisfaction of training the men who had enlisted for regular service with the regiment, but this was now gone. Men for the whole Lancastrian Brigade were now trained at Hadrian's Camp, just outside Carlisle. National Service had long since diminished loyalty to any particular regiment. The army was being rationalised.

The officers at the depot found themselves doing a variety of jobs; part public relations with the local community; part liaison with the territorial battalions and regimental associations; part looking after itinerants awaiting posting or, like myself, about to leave the army.

We travelled to Penrith and Workington when the local regimental associations were having reunion dinners, and an elderly general or brigadier from the regiment would be brought out of retirement to make a speech, his words punctuated by 'Hear-hears'. They were sad, meagre occasions with poor food and a few flowers in some barely disguised municipal restaurant. We would wear our miniature medals on our lapels, admiring the old soldiers' Great War gongs, bigger than pennies, supported on faded ribbon in patriotic colours. The meal would end with a glass of port, served from bottles, and we drank the Loyal Toast before driving back in the cold night air.

In the mess at Carlisle Castle the officers gossiped about their colleagues in the First Battalion, about others who had served with the regiment during World War II, some of whom kept in touch with the regiment from their market gardens or small businesses. Old Colonel Smyth, who had retired years ago, and who was the linchpin of the regimental association, padded about the regimental museum in his blazer and grey flannels, permanently suede-booted.

The regiment had an impressive museum, with uniforms going back to the early eighteenth century when the Thirty-Fourth Regiment of Foot was raised in the border counties. The development of different styles of uniform, head-dress, buttons, and accoutrements for officers, warrant officers and men was traced in the glass show-cases, with examples of each. There were also photographs from 1914, 1918 and the twenties; elongated affairs, mounted on the wall, with a whole battalion shown against grey asphalt and damp billets at Aldershot.

Christmas came, and the major commanding the depot asked me to serve as adjutant and acting OC whilst he and the other officers went off for the Christmas holiday. I had no special plans and so stayed behind whilst the officers and almost all the men went off home for a few days. A cook and a mess servant stayed behind to look after me. I lunched and dined alone in solitude, thinking that this was the first and last time I would have a castle as my bailiwick, if only for a few days. The mess servant banked up the fire in the ante-room after dinner, then asked to go off duty if I required no more drinks. I sat alone before the fire with a brandy, glancing through the back issues of the *Illustrated London News*, the *Tatler*, the *Field*, *Countryman*, and *Blackwood's Magazine*. I would be unlikely to see the familiar weeklies of the mess outside the army.

Life resumed at the depot after a few days. My release was only a few weeks away now. Snow dotted the Cumberland fells above the Penrith road. I took some detachments of recruits in their final week of training from Hadrian's Camp down to the ranges at Warcop, where they fired rifle, Bren, mortar and threw grenades. At noon we ate the doorsteps of sandwiches provided by the cook-house at Hadrian's Camp, swilled pints of sweet tea, and went back to the ranges again to paint the cold gun barrels and the ammunition clips with the five bright prongs of the .303

bullets. At the end of the day the soldiers tugged the soft four by two lint through their gun barrels. I inspected the barrels they held up, peering down at the grubby thumb nails held over the bolt mechanism to reflect the vestige of pale winter light. Then we climbed into the three-tonners and came back to Carlisle, the tyres humming on the icy roads beneath us as the light gave out.

In my final week at Carlisle an NCO's court martial came up at Hadrian's Camp. The OC asked me to serve as defending officer for the accused man, a corporal with about eleven years' service who would soon be up for promotion to sergeant. He was a motor transport corporal, with many vehicles in his care, and he had foolishly taken a small van on several occasions for his private use, using army petrol. The military police had trailed him more than once. He was discovered one night with a girl, not his wife, on a lonely country road.

It was some time since I had studied military law and the duties of defending counsel. I retrieved some Sandhurst notes and studied Army Regulations and the Manual of Military Law for a few days before seeing the NCO. Until his trial he was under open arrest in Hadrian's Camp. He was very depressed about the matter, convinced that his career was ruined. At the very least he expected to be reduced to the ranks and put in prison. I saw his wife, a worried, wan-looking creature, who told me she was prepared to stand by her husband, despite the 'goings on'. Then I saw the corporal again and had a word with the MT officers and some of the warrant officers and sergeants in the MT lines.

The court martial assembled on a Monday at a barracks in Longtown, north of Carlisle, so that the presiding officer, brought up from London, should not come into contact with prosecuting and defending counsel in the mess. The trial lasted two days, and my final speech for the defence came last in the proceedings. I pinned the defence to the corporal's previous good record, the aberration which the present offence clearly represented, and his wife's continued loyalty and determination to rescue their marriage, despite this setback. There were two children to the marriage, and I put it to the tribunal that the corporal's marriage

and the future of his children were in their hands, as much as
his army career. The court recessed, and judgement was given: a
severe reprimand. The corporal would retain his rank, and the
court hoped that he would now put this matter behind him and
forget it. Military law is often thought to be heartless. But military justice inside the army could always be tempered with
compassion.

The corporal came to see me next day. He looked much smarter
and he carried a note from his wife. She said how grateful she
was, and that she was afraid she might embarrass everybody by
crying if she came personally. I was glad this was my last contribution to army life. I left at the end of the same week. The final
letter from the War Office was brief and formal, even down to the
involuntarism of the last sentence, signed by the lieutenant-
general who was military secretary at the time: 'I am to conclude
by taking this opportunity of thanking you for your services while
on the active list'.

The depot company office passed me my last rail warrant, a
single ticket home. I had an hour to fill in before the train, so I
asked Colonel Smyth for a key and walked over to the regimental
museum beneath the castle walls. I wanted to have a last look at
the photographs of the First and Second Battalions at different
times during the past fifty years or so.

The formal photographs on the cold, stone walls of the museum
were characteristically impersonal: officers in front, sitting down,
and six or seven geometrically straight rows of warrant officers
and ORs ranged up behind them. The yard-long battalion photographs were taken by a slow semi-circular sweep of the camera,
whilst everyone stood still. The earlier ones were sepia, yet
remarkably clear.

Some of the gifts to the regimental museum had not yet been
placed in the exhibition cases but lay unsorted in an alcove. I
glanced among them and found an album of photographs, bound
in watered silk. The prints showed informal aspects of officers'
life in India during the late twenties and early thirties. I recognised some of the faces; majors now, in their late forties or fifties,
serving at Tel el Kebir. Others had retired, though the names

beneath the photo were familiar to me from anecdotes in the mess. Some were killed in World War II.

One photograph showed some young officers in India in the early thirties sitting nonchalantly on chairs spread about a lawn. One or two wore riding habit, pith-helmeted, leaning on polo sticks; others were in white ducks, with silk squares neat at the throat. One was in uniform, with sword and Sam Browne belt. Orderly officer of the day, probably. At the side, an Indian mess servant held a tray of drinks. The bright sunshine threw deep shadows from the cane chairs on to the lawn. Possibly the photograph had been taken at high noon on a Sunday. There was no sign of imperial twilight. It was one of the most fascinating historical documents I had met. Twenty years had turned the world about.

Epilogue

The train took me east from Cumberland near the line of Hadrian's Wall and the Northumberland moors where I had spent much of my boyhood. It was just over five years since I had travelled west to Belfast. I wondered if it had all been worthwhile. It was too early to say. Another ten or twenty years perhaps, then I would know.

I got back to my books. Spring came and I travelled down to Oxford to take the entrance examination at Merton, University, and St John's colleges. The Merton examinations came first, spread over two days. I enjoyed the papers and the interview with the Fellows seemed congenial.

I set out for the University College examinations next morning, but the Merton chaplain greeted me in Merton Street. I remembered him from the interview. He told me that if Merton was my first choice I needn't take the other examinations. Merton would be letting me know that I had a place, though it was not formally declared yet. It was a kind gesture, typical of Stephen Williams. I withdrew from the other examinations. Merton it was, then, and I was glad. A misty, metaphysical college, slightly apart from the mainstream of Oxford, its chimes all out of joint. Max Beerbohm and T. S. Eliot, Edmund Blunden and Louis MacNeice had been here: men of letters. I hoped some of their spirit might rub off.

Summer passed and I put away military things in the most literal sense. A deep carton took all of them, including the journal. I tied the box with a rope, scribbled on the top with a red crayon '5 & 7', and put it away. It gathered dust. I went up to Oxford as the years of austerity were shading into the age of affluence. Looking back now, I can see that my years with the military made me

enjoy university infinitely more than a progression from school
would have allowed. One had learnt to organise one's day. After
the spartan life, student life was almost sybaritic. I had worked
hard at school, but intellectual constipation set in. I wanted to
meet the real world to which the books referred, if only to see
whether it was real, and what were the human dimensions of the
bookish world. Having seen it, or a part of it, I was content, ready
to get back to speculation, ideas, and theories. Looking back, I
would do it again: without a shadow of a doubt I would do it
again.

In the fifties, beyond the academy's cloistered walls, the retreat
from empire continued and the British army took hard knocks.
In 1959 a newspaper item I noticed in *The Times* mentioned
Army Order no 40, by command of Elizabeth Regina. The word-
ing was terse, unsentimental.

> Whereas we deem it expedient to make certain changes in the com-
> position of our military forces and in the designation of certain infantry
> regiments therein; our Will and Pleasure is that the following regiments
> will be amalgamated, and the amalgated regiment designated as under...

I glanced down the list. The Border Regiment was to lose its
identity, merging with the King's Own Royal Regiment to form
the King's Own Border Regiment. I wondered what further bleak
little ceremony would be performed on some asphalt square
between Lancaster and Carlisle.

In 1969, another ten years later, county regiments were axed
again. Rationalisation was the cry. This is not to say that today's
smaller army is not intensely professional and fully adequate to
its tasks. It manifestly is, and man for man it is probably a more
efficient fighting machine, equipped with much better weapons
and with greater firing power. But I wondered if the rationalisers
really grasped that the fundamental problem today may be the
sense of belonging. One cannot *belong* to a logistical matrix.
Glancing at the Army List, I found few of the names I knew at
Sandhurst. Most had wanted to join their county regiment: they
were not ashamed to say so. Many would have retired early,
denied the promotion they had good reason to expect when they

entered the profession with high hopes. Some I knew were dead. Those that remained were largely taken for granted during long years of peace. Yet peace is a fragile commodity, and democracies take their armies for granted more than dictatorships are inclined to do. That is their weakness and their strength.

For myself, I recognised a private debt, chiefly to men older than myself, who tried to make a soldier out of me. If they did not quite succeed, nevertheless there was a debt, difficult to convey precisely. But John Mulgan expressed it well in his book, speaking across the grave. With him I would agree:

> We have cause enough to be grateful to the old soldiers who held the Army together and taught us how to live and carry out orders. Most of them brought their knowledge from long years abroad in dust-dried barracks. There was some fragment of truth behind the Kipling legend. None of them had had much fun. They had led a life without the normal compensations that a man gets by making something or by working intelligently. Behind them lay the eternal gossip of military communities, the struggle to have children and a home, a future of small pensions and menial jobs in civil life. But the bitterness and the system produced an austerity which is useful for war, and you thanked God when you met them—the straight-backed, humourless, reliable men, without imagination but also without temperament, who do what they say they will do, and often get killed in the process.